About the Wellcome Trust

The Wellcome Trust is an independent charity whose mission is to foster and promote research with the aim of improving human and animal health. We have three principal aims:

1. *Advancing knowledge:* to support research to increase understanding of health and disease, and its societal context
2. *Using knowledge:* to support the development and use of knowledge to create health benefit
3. *Engaging society:* to engage with society to foster an informed climate within which biomedical research can flourish.

In support of these aims, we also recognise the importance of promoting the development of individuals we fund, enhancing the environment for research and its application, and constantly improving the way we operate.

About Demos

Who we are

Demos is the think tank for everyday democracy. We believe everyone should be able to make personal choices in their daily lives that contribute to the common good. Our aim is to put this democratic idea into practice by working with organisations in ways that make them more effective and legitimate.

What we work on

We focus on six areas: public services; science and technology; cities and public space; people and communities; arts and culture; and global security.

Who we work with

Our partners include policy-makers, companies, public service providers and social entrepreneurs. Demos is not linked to any party but we work with politicians across political divides. Our international network – which extends across Eastern Europe, Scandinavia, Australia, Brazil, India and China – provides a global perspective and enables us to work across borders.

How we work

Demos knows the importance of learning from experience. We test and improve our ideas in practice by working with people who can make change happen. Our collaborative approach means that our partners share in the creation and ownership of new ideas.

What we offer

We analyse social and political change, which we connect to innovation and learning in organisations. We help our partners show thought leadership and respond to emerging policy challenges.

How we communicate

As an independent voice, we can create debates that lead to real change. We use the media, public events, workshops and publications to communicate our ideas. All our books can be downloaded free from the Demos website.

www.demos.co.uk

First published in 2006

ISBN 1 84180 155 0
Copy edited by Julie Pickard
Typeset by utimestwo, Collingtree, Northants
Printed by Calverts, London

For further information and
subscription details please contact:

Demos
Magdalen House
136 Tooley Street
London SE1 2TU

telephone: 0845 458 5949
email: hello@demos.co.uk
web: www.demos.co.uk

Better Humans?

The politics of human enhancement and life extension

Edited by
Paul Miller
James Wilsdon

creative
commons

Contents

Acknowledgements

Many thanks to the Wellcome Trust, our partner and funder for this Demos collection. We are particularly grateful to Sarah Hornby, Caroline Hurren and Clare Matterson at Wellcome for their support and guidance. It is important to note that the Wellcome Trust does not necessarily endorse the views expressed in all the essays – but they are to be applauded for seeking to stimulate a wider debate of this important topic.

Thanks also to the public participants of the Cardiff and Nottingham Café Scientifiques, who influenced our thinking by sharing their hopes and fears about human enhancement – and particularly to Rachel Dodds and Melanie Heeley who organised those sessions. Thanks also to Steve Rayner and Peter Healey at the James Martin Institute and Claire Craig and Andrew Jackson at Foresight for their advice on themes and contributors.

Several members of the Demos team made useful contributions: Tom Bentley, Sam Hinton-Smith, Abi Hewitt, Alison Harvie, Julia Huber and Jack Stilgoe. Thanks also to Julie Pickard for her tireless copy-editing.

Last, but by no means least, our thanks to all the authors for their thought-provoking essays, and for supporting the idea of a collection on this theme.

Paul Miller and James Wilsdon
January 2006

Foreword

Mark Walport

Some of the most exciting scientific developments in recent times have been achieved through advances in biomedical research. The impact of these advances on human health is enormous and the potential benefits far reaching. For example, it is not so long ago that the complete sequencing of the human genome seemed an improbable challenge. We now know the human genome sequence and this is likely to prove a landmark in our understanding of health and disease as well as providing a greater understanding of our own diversity.

Science and technology is constantly extending our knowledge of what it means to be human and our relationship with other species. However, exciting new scientific discoveries also carry with them ethical and moral dilemmas. For example, the use of stem cell technology in treating disease may have huge potential, yet it also raises important questions about the value of its use for health gain. In these instances when advances in science and technology provide important ethical challenges, it is crucial that society engages in discussion and maintains confidence in the process of scientific discovery. This volume contains a series of stimulating and in some cases, controversial essays on human enhancement. They may not depict real futures but are a provocative basis for debate.

The Wellcome Trust is committed to engaging with society on the interests and concerns raised by the biomedical sciences past, present and future. It is only by doing this that we can balance the needs of the research endeavour with those of society and foster an informed climate within which biomedical research can flourish.

Mark Walport is Director of the Wellcome Trust.

1. Stronger, longer, smarter, faster

Paul Miller and James Wilsdon

In a packed bar in Nottingham, the temperature of the conversation is rising. We've been talking about the Flynn effect, the fact that human intelligence – as measured by IQ tests when they aren't re-based for each generation – has been steadily improving for the past 100 years. So much so that someone who scored in the top 10 per cent in 1900 would only make it into the bottom 5 per cent in 2000. This is Nottingham's Café Scientifique, one of the most vibrant in the country, where each Monday evening people meet to drink, talk and argue about science.

We go on to discuss drugs that affect the brain. Ritalin (methylphenidate) and other stimulants are now being prescribed to between four and five million school students in the US.[1] Others are borrowing, buying or selling the tablets and taking them to boost their concentration, especially in the run-up to exams. Does this mean they are cheating? And what about the next generation of drugs emerging from the laboratory, which will improve our memories, or overcome the need for sleep? How widespread will the use of such 'Viagra for the brain' become?

Opinion in the bar is divided. 'What's your definition of intelligence?' asks one person.

'I'm not sure I want my memory improving – there are some things you forget for a reason,' says another.

Some see a difference between enhancement technologies that are available now and those that might alter future generations in more fundamental ways. We talk about the radical end of the enhancement spectrum – the possibility of changing the genetic make-up of our children, of inserting artificial implants into our bodies, or of 'uploading' our brains into a new virtual form. 'Maybe natural selection doesn't work any more,' says one young woman. 'Maybe we need to artificially make ourselves more suited to our environment because genetics can't keep up.' 'Who are we to say that?' counters a middle-aged man. 'How do we know what characteristics are going to be useful in the future?'

Finally, we talk about the possibilities of radical life extension – of living to 150 or beyond. We describe the predictions of Aubrey de Grey, the Cambridge scientist, who argues that there is no reason why people shouldn't eventually live to 1000. It's not a prospect that appeals to everybody. 'It's like that Queen song,' says someone. 'Who wants to live forever? Not me.'

As the crowd in the bar starts to disperse, the sense is that we've only scraped the surface of some pretty complex debates. The appetite for information is greater than we, as the facilitators of the session, can satisfy.

Better definitions

This collection explores the next generation of technologies for human enhancement, and what they might mean for society. Definitions of enhancement vary, but the term usually refers to interventions designed to improve human performance beyond what is required to sustain or restore good health.[2] In the title of one recent book, enhancements aim to make us 'better than well'.[3]

We all share a desire for self-improvement. Whether through education, work, parenthood or adhering to religious or ethical codes, each of us seeks to become a 'better human' in a variety of ways. And for many people, more consumerist pursuits hold the key to self-improvement: working out in the gym, wearing makeup, buying new clothes or indulging in a spot of cosmetic surgery.

The starting point for this collection is that a new set of

possibilities for enhancement is now opening up. Advances in biotechnology, neuroscience, computing, pharmacology and nanotechnology mean that we are in the early stages of a new period of human technological potential. Some of these possibilities are already with us; others remain the preserve of science fiction. Table 1 summarises the current status of the main types of enhancement technology.

As these technologies develop and are shaped by parallel changes in culture and consumer expectation, we may well see a surge in demand for enhancements – surgical, chemical, robotic, genetic – which cannot be categorised as 'medical' but which strengthen our mental or physical capabilities. Some types of enhancement will progress incrementally – for example, new 'smart' drugs – while others are likely to prove more disruptive – for example, nanotechnology or gene therapy.

The question that this collection tries to answer is 'will such enhancement technologies make things better?' not only in terms of human performance but also in terms of our collective well-being and quality of life. As enhancements become more widely available, they will inevitably prompt debate about the limits of their use, and whether they can and should be regulated. At a deeper level, they also force us to address questions of identity, personhood, responsibility and democracy, and about the long-term consequences of altering human nature and capabilities.

The collection is divided into two parts. In the first part, three prominent advocates of enhancement set out their case. Arthur Caplan insists that the new enhancement technologies are merely the logical next step in an ongoing process of using new knowledge to improve ourselves. He criticises the 'anti-meliorists', who argue for 'a distinct essence, a kind of template of humanity that somehow is in there as a core that cannot be touched or changed or manipulated'. On the contrary, he says, 'I find no in-principle arguments why we shouldn't try to improve ourselves at all. I don't find it persuasive that to say you want to be stronger, faster, smarter makes you vain. . . . That's what agriculture is. That's what plumbing is. That's what

Table 1. Current status of the main types of enhancement technology

Name	Function	Examples	Availability
Psychopharma-cology	Alteration of brain state or mood	Prozac (enhanced mood), Ritalin (enhances concentration), Provigil (prolongs alert wakefulness)	Now available; other research in progress, eg appetite suppressants
Other pharma-cological agents	Alteration of bodily form or function	Growth hormone, Viagra (male sexual function), erythropoietin (athletic perfor-mance), steroids (muscle mass)	All now available
Cosmetic surgery	Changes to facial or physical appearance		Now widely available
Preimplantation genetic diagnosis	Enables embryos to be selected for particular genetic traits	Huntingdon's disease, cystic fibrosis[4]	Available for several dozen illnesses; more genetic tests being developed
Gene therapy	Alters genetic make-up of selected cells in the body	Somatic therapy – various experimental treatments	Germline therapy – GM plants; mouse embryo engineering Somatic gene therapy is being used in a number of experi-mental treatments. Human germline gene therapy is currently illegal in the UK, although the House of Commons Science and Technology Committee has recommended it should be permitted for research purposes[5]

Name	Function	Examples	Availability
Cybernetics	Alteration of mental or physical function by embedding engineering or electronic systems within the body	Kevin Warwick's research at Reading University on human–computer interactions[6]	Research actively being pursued
Nanotechnologies	Similar to cybernetics, but using far higher levels of miniaturisation	Nanodevices to destroy tumours or rebuild cell walls	According to the UK's Royal Society, at least 10 years away[7]
Radical life extension	Combination of techniques enabling human lifespans to reach 150 years or more	Theoretical possibilities hotly debated by scientists such as Aubrey de Grey and S Jay Olshansky	Even the most optimistic predictions (de Grey) suggest we are 25–30 years away from the necessary scientific breakthroughs

clothes are. That's what transportation systems are. They are all attempts by us to transcend our nature. Do they make us less human?'

Nick Bostrom goes further still, arguing that the hardware, software and input/output mechanisms required for 'posthuman' forms of artificial intelligence will be available within 50 years. At this point, we may reach what has been dubbed the 'singularity' – 'a hypothetical point in the future where the rate of technological progress becomes so rapid that the world is radically transformed virtually overnight. . . . Superintelligent machines would then be able to rapidly advance all other fields of science and technology. Among the many other things that would become possible is the uploading of human minds into computers, and dramatic modification or enhancement of the biological capacities of human beings that remain organic.'

And we meet Aubrey de Grey, the self-confessed 'crusader' for human longevity, who suggests that 'we will inevitably be able to address ageing just as effectively as we address many diseases today. . . . I think the first person to live to 1000 might be 60 already.' de Grey argues that society is caught in a 'pro-ageing trance' which leads most of us to defend 'the indefinite perpetuation of what it is in fact humanity's primary duty to eliminate as soon as possible'. He believes that saving a life by deferring unnecessary death is a moral imperative equivalent to that of providing development aid to prevent a child dying from malnutrition.

The world's most dangerous idea?

In the middle of the stage in a darkened lecture hall in Stanford University, all eyes are fixed on a black box, roughly six feet tall and three feet wide. We are at the 2003 Accelerating Change conference – a gathering of the West Coast digerati. And we are waiting for Ray Kurzweil, the celebrated inventor and futurist, to appear before us in 3D holographic form.

Kurzweil is one of the intellectual figureheads of a movement which has come to be known as transhumanism. According to the World Transhumanist Association, this is the belief that 'the human species in its current form does not represent the end of our development but rather a comparatively early phase'.[8] In his most recent book, *The Singularity is Near*, Kurzweil argues that the scope for enhancement will follow an exponential curve, rather than a linear trend. 'Ultimately we will merge with our technology. . . . By the mid 2040s, the non-biological portion of our intelligence will be billions of times more capable than the biological portion.'[9]

In the lecture hall, the audience is growing impatient. We keep expecting Kurzweil to shimmer into view, rather like the projection of Princess Leia from R2D2 in *Star Wars*. But in the end, he never appeared. On this occasion, the technology hadn't accelerated quite enough. In fact it just didn't work. We could hear Kurzweil (well, four words out of five anyway) and we could see his PowerPoint slides (although they kept on crashing), but try as the army of engineers present did, we still couldn't see him sitting in his office in Boston. In the centre of the stage, the huge piece of kit that was supposed to represent the future stood dormant and useless.

Yet despite the occasional glitch, as Greg Klerkx describes in his essay, the past few years have seen a surge of support for transhumanism. The current generation of thinkers represent 'what might be called transhumanism's third wave'. They are 'sounding a clarion call that radically improved and longer-lived humans are imminent, and they are basing such claims on optimistic extrapolations from relatively new science and technology'. And as the column inches devoted to Kurzweil, de Grey and others demonstrate, these ideas and their charismatic protagonists fascinate the media.

It's easy to laugh at or dismiss the transhumanists as eccentric cranks, inhabiting the outer margins of science. But some serious commentators are ringing alarm bells. Francis Fukuyama, Professor of International Political Economy at Johns Hopkins University, has called transhumanism 'the world's most dangerous idea'. In a 2004 article for *Foreign Policy* he warned: 'Society is unlikely to fall suddenly under the spell of the transhumanist worldview. But it is very possible that we will nibble at biotechnology's tempting offerings without realizing that they come at a frightful moral cost.'[10] Fukuyama is a member of the influential President's Council on Bioethics, which George W Bush set up in 2001. In a series of reports, the Council has advocated a conservative position on enhancement, stem cell research and human cloning, prompting heated responses from the transhumanists.

But it is not only Republican conservatives who feel a sense of unease about enhancement. The second part of this collection features contributions from a range of scientists, social scientists and writers, who raise questions and concerns about the potential implications of these developments. Steven Rose, the prominent neuroscientist, anticipates many positive benefits flowing from advances in his field. But he worries that 'there will also be attempts to develop physical techniques for altering mental processes. These include techniques for direct surveillance of citizen's thoughts, which could be used for pre-emptive incarceration or medical treatment.'

Danielle Turner and Barbara Sahakian, neuropsychologists at Cambridge University, focus on the effects that smart drugs might have on children and students: 'Is it possible that these drugs could be used to reduce social inequality and injustice in society? Or it is more likely that their use will fuel further disparity based on a lack of affordability? Could cognitive enhancers have unexpected social ramifications, as people are deprived of a sense of satisfaction at their own achievements?'

Sarah Franklin addresses the beginning of life in her analysis of the public debate that surrounds preimplantation genetic diagnosis (or PGD) – which in the media's eyes tends to be equated with 'designer babies'. Like the figure of the human clone, 'the designer baby has become an iconic signifier of the dilemmas and risks posed by new genetic technologies'. But despite fears that we will forfeit our humanity to such advances, Franklin is optimistic about the prospects for genuine deliberation: 'What emerges from a brief scan of PGD and its future, is the extent to which [it] is associated with public debate and regulation, *not their absence*.'

For Jon Turney, it is the prospect of death that raises the deepest questions. He surveys four decades of literary speculation about immortality, and is forced to conclude that: 'a search for immortality seems to me a counsel of despair, not hope. As completely unlimited life is out of the question, what is the appeal of staking all on such a fantasy? If a life limited to 100 years is devoid of meaning, why would living to 200, or even 2000, improve matters? There would still be infinitely many years of non-being to follow.'

Decca Aitkenhead writes about the one form of enhancement that is already booming: cosmetic surgery. 'Cosmetic operations in BUPA hospitals were up by 32 per cent last year, male patient numbers more than doubled, and operations by the British Association of Aesthetic Plastic Surgeons (BAAPS) rose by 50 per cent.' Something that was once regarded as shameful or taboo has been rebranded as mundane, and popularised through TV makeover shows and magazine competitions where the prize is breast augmentation.

The most critical voices in the collection are those of the disability rights campaigners, Rachel Hurst and Gregor Wolbring. In her powerful essay, Hurst argues that the techniques and motivations for human enhancement are akin to eugenics: 'We will never be able to continue building a society based on human rights while genetic advances are directed towards the elimination of disabling impairment. The most important right – the right to life itself – can never be ensured in this climate.' For Wolbring, there is a danger that 'the transhumanist model sees every human body as defective and in need of improvement, such that every unenhanced human being is, by definition, "disabled" in the impairment or medical sense'. This will give rise to a new, unenhanced underclass.

Better democracy

Each of the essays grapples in its own way with the crucial question 'Who should decide?' For most of the authors, as for Demos, the starting point is a commitment to democracy. Yet in order to exercise any democratic oversight of new forms of enhancement, we simultaneously need to 'enhance' our ability to make choices about what we value in our lives. And we need to recognise that being human depends far more on our capability for engaging in meaningful forms of collective deliberation than on any new technology or advance in processing power. This is a point made well by Raj Persaud in his essay, where he points out that enhancement 'requires us not to become different in order to improve, but rather to become more like the good parts of ourselves. Enhanced people are already walking around among us, but we tend to ignore them. We do this at our peril and new technologies will not save us from this mistake.'

This crucial point is often lost in the deterministic predictions of the transhumanists. All too easily, they slide from a discussion of what new technologies *may* make possible to an assumption that these changes *will* happen, without any appreciation of the subtleties of culture and values, or the unpredictable twists and turns of democracy. Dan Sarewitz, the sociologist of science, describes how he took part in a number of National Science Foundation meetings on human enhancement, where these limitations were apparent.

> *Most of the attendees were highly intelligent white males who worked in the semiconductor industry, at national weapons laboratories or major research universities. At one point, the group got to talking about how we might soon achieve brain-to-brain interfaces that would eliminate misunderstandings among humans. Instead of having to rely on imperfect words, we would be able to directly signal our thoughts with perfect precision.*
> *I asked how such enhanced abilities would get around differing values and interests. For instance, how would more direct communication of thought help Israelis and Palestinians better understand one another? Unable to use the ambiguities and subtleties of language to soften the impact of one's raw convictions, might conflict actually be amplified? A person at one of the meetings acknowledged he 'hadn't thought about values', while another suggested that I was being overly negative. . . . This sort of conceptual cluelessness is rampant in the world of techno-optimism.*[11]

Part of the problem stems from transhumanism's origins in a particular strain of Silicon Valley libertarianism (an ideology described with amusing candour by former *Wired* writer Paulina Borsook in her book *Cyberselfish*[12]). Yet there are now efforts under way from within the transhumanist movement to grapple more seriously with these social and political challenges. One of the most interesting books to emerge recently is *Citizen Cyborg* by James Hughes, executive director of the World Transhumanist Association. Hughes argues that enhancement must go hand in hand with a radically strengthened democracy: 'We can embrace the transhuman technologies while proposing democratic ways to manage them and reduce their risks. . . . We need a democratic transhumanist movement fighting both for our right to control our bodies with technology, and for the democratic control, regulation and equitable distribution of those technologies.'[13] He even suggests that transhumanism will become the next progressive force, picking up the mantle of human liberation from the movements for gender and racial equality.

Despite this, Hughes occasionally slips back into the familiar mantras of technohype and determinism. And his call for a less polarised debate isn't helped by him labelling all critics of enhancement 'BioLuddites', without making the effort to engage seriously with the substance, texture and motivation of their concerns. Nonetheless, his book is a welcome contribution, particularly as it has sparked a great deal of debate within the transhumanist movement itself.

Better policy

Hughes also points to where discussions about enhancement need to go next. The transhumanist movement, insofar as it exists as a defined community, can no longer own or manage the terms of these debates. These technologies have the potential to affect all of us, and they must now be opened up to wider processes of democratic scrutiny and debate. In particular, there needs to be a distinctively European discussion of these issues, as opposed to a wholesale import of debates from the US, where the religious right tends to set the terms of critical discussion. Is it possible to look afresh at some of the social and ethical dilemmas raised by enhancement from a more progressive European stance? A forthcoming conference organised by Oxford University's James Martin Institute is a positive step in this direction.[14]

This links to the wider question of how we improve our social readiness and the resilience of our systems of governance to cope with these changes. We close with three practical suggestions.

1. Upstream public engagement

As Demos has argued elsewhere, there is a need to move public engagement 'upstream', to an earlier stage in processes of research and development.[15] A number of experiments are now under way in this regard in the UK, born out of recognition that earlier controversies, such as those around genetically modified crops, have created a window of opportunity to improve the governance of science and technology.[16]

Enhancement technologies are an area where dialogue is urgently required between scientists, policy-makers, bioethicists, healthcare professionals, educationalists, NGOs, disability groups and the wider public. Such discussions should address not only narrowly framed 'impacts', but also the wider social and ethical context in which such innovations may occur, for example, how to define the benefits of different forms of enhancement in terms of well-being and life satisfaction. Or how to determine what constitutes a good death, as well as an enhanced life. This dialogue should be facilitated by key players within government and the scientific community, such as the Office of Science and Technology and the Royal Society. The UK government should also look seriously at the option of establishing a Commission on Emerging Technologies and Society, which could provide an institutional hub for 'public engagement and social assessment of technologies'.[17]

2. The new old

If even a handful of the predictions of the transhumanists are accurate, we face the prospect of life expectancy in our already ageing society rising far more dramatically than current models suggest. By and large, this will be a very positive development – potentially allowing more people to live fuller and healthier lives into their 80s, 90s and beyond (far beyond if you accept the arguments of Aubrey de Grey!).

But it will also create some challenges. In the UK, one only has to look at the fierce response from some quarters to the proposal by Lord Turner's Pensions Commission to raise the retirement age from 65 to 69 over the next 30 or 40 years to see just how far we have to go if we are to face up to the potential for a life-extended society.[18] A much steeper retirement 'escalator' may be required, with a retirement age of 80 or 90 becoming necessary well within the lifetime of Lord Turner's proposals.

As these proposals are now the subject of further consultation and debate, we suggest that further analysis is carried out by HM Treasury and the Office of Science and Technology about the potential implications of more radical forms of enhancement and life extension.

3. Education epidemic

Finally, there is a need to look seriously at the implications of enhancement for our education system. On the positive side, advances in neuroscience mean that we are developing a more sophisticated understanding of how young people's brains develop and learn, and this knowledge can inform educational policy and practice.

More negatively, the widespread use of Ritalin and the potential for new types of pharmacological enhancement threaten to undermine systems of fair assessment. The response to this should take two forms. First, it requires us to rethink the role of competitive exams in our education system, which are likely to encourage the use of cognitive enhancers, and instead place greater emphasis on individual learning pathways that equip students for a lifetime of learning.[19]

Second, efforts to restrict recreational drug use in schools, for example through random drug testing, will need to broaden their scope to cope with a new generation of drugs, whose educational impact is potentially far more significant than drugs such as cannabis or Ecstasy. Whereas recreational drugs tend to be taken without the support of parents and teachers, we face the prospect of enhancement drugs being actively 'pushed' to under-performing students by teachers or parents. Just as the scandal of drugs in sport led to the creation of the World Anti-Doping Agency in 1999, with its motto 'play true', so the government should consider creating a schools and universities anti-doping agency (with the motto 'learn true') to promote a drugs-free education system.

The most important thing we can do when confronted with the new possibilities for human enhancement is to get people talking. We may not have accurate foresight but we can have forethought. And scientists or self-proclaimed transhumanists cannot retreat into their own neatly defined boxes. Rather, they have to roll up their sleeves and get stuck in to a meaningful dialogue with citizens and policy-makers about what might happen, and how those trajectories can still be shaped and changed.

The aim of this collection is just that: to encourage a wider debate before these technologies are a done deal. And to start more of the kinds of discussions that took place in that bar in Nottingham.

Paul Miller is a Demos associate and James Wilsdon is head of science and innovation at Demos.

Notes

1 L Diller, *Running on Ritalin: A physician reflects on children, society, and performance in a pill* (New York: Bantam, 1999).
2 See, eg E Juengst, 'What does "enhancement" mean?' in E Parens (ed), *Enhancing Human Traits: Ethical and social implications* (Washington DC: Georgetown University Press, 1998).
3 C Elliott, *Better Than Well: American medicine meets the American dream* (New York: W Norton & Co, 2003).
4 The Human Fertilisation and Embryology Authority website includes a list of diseases for which PGD is licensed in the UK; see: www.hfea.gov.uk/Home (accessed 12 Jan 2006).
5 House of Commons Science and Technology Committee, *Fifth Report: Human reproductive technologies and the law* (London: House of Commons, March 2005).
6 See www.cyber.reading.ac.uk/people/K.Warwick.htm (accessed 12 Jan 2006).
7 Royal Society and Royal Academy of Engineering Nanoscience and Nanotechnologies: opportunities and uncertainties (London: Royal Society/RAE, July 2004).
8 http://transhumanism.org/index.php/WTA/faq/ (accessed 12 Jan 2006).
9 R Kurzweil, *The Singularity is Near: When humans transcend biology* (New York: Viking, 2005).
10 F Fukuyama, 'Transhumanism', *Foreign Policy*, Sep/Oct 2004.
11 D Sarewitz, 'Will enhancement make us better?', *Los Angeles Times*, 9 Aug 2005.
12 P Borsook, *Cyberselfish: A critical romp through the world of high-tech* (London: Little Brown, 2000).
13 J Hughes, *Citizen Cyborg* (Cambridge, MA: Westview Press, 2004).
14 See www.martininstitute.ox.ac.uk/jmi/forum2006/ (accessed 12 Jan 2006).
15 J Wilsdon and R Willis, *See-through Science: Why public engagement needs to move upstream* (London: Demos, 2004); J Wilsdon, B Wynne and J Stilgoe, *The Public Value of Science* (London: Demos, 2005).

16 For example, see www.demos.co.uk/projects/currentprojects/nanodialogues/ (accessed 12 Jan 2006).
17 For more on this proposal, see Wilsdon et al, *The Public Value of Science.*
18 See www.pensionscommission.org.uk (accessed 12 Jan 2006).
19 For more on this, see P Skidmore, *Beyond Measure: Why educational assessment is failing the test* (London: Demos, 2003).

Part 1: The case for enhancement

2. Is it wrong to try to improve human nature?

Arthur Caplan

I walked by a laser eye surgery clinic in a shopping mall recently and on the door it said, 'Not only is this procedure easy and painless and quick, you will see better than 20/20.' They were saying that they will make you see better than the best nature can provide. There are very few people who have vision that good. On the whole, few of us see 20/20 and almost none of us sees better than that. And what the clinic was saying was 'We can make you see better than ever before, with your contacts, with what biology gave you, with what your glasses could provide. This is going to work better.' Now, is that immoral?

If I go to the laser surgery place and have my eyes tweaked, and I come out with better than what the limits of biology designed into me, 20/20 vision if I was lucky, am I committing a moral wrong? Am I vain? Is it inequitable because other people don't get their eyeballs done or couldn't afford it?

Is it something that we have to say is inauthentic? I didn't really earn it. I guess I didn't exercise my eyes. I didn't try to avoid staring too much at a computer screen. I didn't do the things that might have helped my vision along. I just lay down, the laser did its thing and my eyes are seeing better than ever.

And going even further, is it a distortion of who we are? If all of us run around with 20/15 vision, are we less than human? Have we become some sort of bizarre, unrecognisable, different type of being, disconnected from who we are today, unrecognisable to our forbears because we see better than any of them ever could have?

Is, ought or can?

A constant interest of mine over the years has been evolutionary biology. The first work of mine that got published was something called 'The Sociobiology Debate', and it was concerned with attempts to infer claims about what was natural or right or good from assertions about biology. Sociobiology certainly was controversial in its day. Although, interestingly enough, it too has gone on to become a standard part of the sciences.

That has led a lot of people over the years to try to draw inferences about what biology tells us about ethics. I believe that it is possible to make some inferences from our biology about certain normative claims or ethical – that is, value-laden – claims about what health and disease are. But in one sense, this sounds like you're violating a principle that philosophers since David Hume have been very stern about. And that principle is: you can't get from facts to values or 'You can't go from an "is" to an "ought"'. This is sometimes referred to as the 'naturalistic fallacy'.

I think it's true that you can't get from 'is' to 'ought'. There is plenty to confirm the long-standing belief that biology and Darwin's theory don't really tell us anything about the limits on what we ought to be, how we ought to behave, what nature we should have, how we ought to design ourselves. That doesn't mean we couldn't come up with some values and principles and arguments that set limits, but I don't think we're going to get them out of biology.

Evolution and biology do tell us things about the limits of what you can do. If somebody says can you put a kidney from person A into person B and their blood types are different – well, biology doesn't care where you put the kidney. But biology cares a lot if you're trying to insert a kidney into a person with a different type of immune system. It tells us that we had better not do it, because the kidney will be rejected.

This is the principle of 'ought' implies 'can'. In order to know what you might want to do, could do, should do, ought to do, you do have to have some idea about what you actually can do. And that's the way I see biology as relevant to thinking about ethics. It tells us what limits exist in nature – regardless of whether we want to overcome the limits, change the limits, change the game, change the rules. I think, ever since Darwin, we haven't had any basis for saying that there's any biological limit on what we could be, should be or might want to become.

The quest for perfection?

There has been a lot of interest on the part of the President's Council on Bioethics in thinking about any number of problems that have arisen with respect to public policy. One of the claims that the President's Council has been wrestling with is the issue of improvement and enhancement. Should we try to improve or enhance ourselves using new biological knowledge?

The council's report *Beyond Therapy*[1] wrestles with the question of what we are going to do with the explosion of knowledge about the brain – some biochemical (drugs that affect the brain); some technological (implants that might go into the brain); some scanning and diagnostic (what are we going to do if we can see the brain and start to make forecasts about propensities or abilities?). What should we do in the face of this new arena of knowledge?

The council isn't alone in having these worries about the wisdom of how we're going to use this new knowledge of the brain to possibly change ourselves. There are other people who have been writing about it, for example Michael Sandel who wrote an article in *The Atlantic Monthly* called 'The case against perfection'.[2] If you start to review these writers' work, you find a number of common themes about what makes them nervous concerning the idea of engineering our nature.

One of the things that's a little unfair about these arguments is that most of the critics are saying, 'You shouldn't pursue perfection.' But, as Salvador Dali said, 'You don't have to worry about perfection; you're not going to get there.' What we're talking about is something different, something more interesting, but it's a little less spooky and that is improvement. 'Should we improve human nature?' is really the question. Not, 'Should we pursue perfection?' I think that rhetoric is an easy mark, an easy point of attack but I want to get it out of the way.

What the anti-meliorists, the anti-improvement people, are trying to argue is, 'Let's not head down the melioristic road.' And they look around and they say, 'You know what's going on right now? Breasts are being augmented. Wrinkles are being smoothed out. Fat is being suctioned out. Blood is being doped and moods are being calmed. If we don't put a stop to this, who knows where we're going to be? Everybody's going to have a breast job. Everybody's going to have pectoral implants. Everybody will run around trying to take drugs to alter their moods – to make them happy or complacent. We have to get on top of this push within the bioengineering side of things to try to change us because it's going to lead to places that we would find unappealing.'

Improvement and vanity

The critics of human enhancement bring forward the argument that, 'If you want to look better, you're vain.' I would have thought, 'If you want to look better, you might say you have self-regard.' You might say that you are trying, in some sense, to present yourself in the world in a way that makes you feel better. You might say that it shows an appropriate level of interest in how others see you. You might even say, if you were sociobiological about it, that it might give you an advantage in the mating game.

But it doesn't just have to be a matter of vanity. If it's really all vain, then why don't we just take off our clothes, throw away the makeup, get rid of the fashion industry and reconcile ourselves to grubbing around in some sort of grass skirt and be done with it? We know that, to some extent, part of what gives us pleasure is trying to control our appearance, control how others see us. It may or may not be something that we can overindulge. I would grant that the person who undergoes perhaps their twentieth cosmetic surgery operation (I have a certain singer in mind here these days) may be abusing the idea of biological change. But that doesn't show that it's always wrong if you don't like the shape of your nose, if you want to remove a port wine stain from your face, if you want to see better through laser surgery, if you'd prefer to wear contacts rather than glasses, if you even want to remove wrinkles. I don't find anything inherently and obviously and self-evidently wrong about this.

There's a part of me that thinks that, when the President's Council gets going on this vanity theme, they're channelling our Protestant ancestors. They're sort of looking to the Puritans and saying, 'Well, my goodness! If you're not praying and laboring, if you have any time to put on deodorant, then what kind of person are you?'

I can also, in defence of some anti-vanity arguments, point out that, if you go to other cultures, say Brazil, they don't have the same hang-ups we do about who's had a facelift or who's had abrasion to remove wrinkles. They say 'If these parts wear out, you fix them. What's your hang-up?' They don't work themselves into a kind of *People Magazine* frenzy and say, 'Well, he had his crow's feet removed and she absolutely had her breasts augmented.'

I'm not arguing that it's right for 14-year-olds to get breast augmentation surgery as a gift, which some have. I think you should learn to decide whether you like your body or not and you're not ready at that age to make such a decision. But again, I'm going to say it's not self-evident to me that all pursuit of beauty or looks or appearance is vain, in and of itself. And certainly vanity has nothing to do with interest in trying to think faster, have more memory, or in the decision about whether one wants to be stronger or to be able to increase aptitudes and capabilities. That's not vanity; that's function.

Equity and fairness

It is true that we could find ourselves, in the developed world, having access to genetic engineering, biological engineering, brain implants, biochemical interventions that poor people in other places cannot get. It's also true that we could find ourselves, within rich countries, with a lot of people unable to buy or purchase many of these things that might enhance or improve capacity.

But I have a very simple question. Is the problem modifying and improving our biological nature? Or is it a problem of inequity?

I'm not in favour of inequity. But, if I said, 'We're going to guarantee access to anyone who wants it to a chip that might be put into somebody's head and improve their memory,' and if I took equity off the table, there's no argument here other than it's bad to have inequity. Inequity is bad. But it's not connected necessarily to biological changes. It's connected to all sorts of important resources.

We already have a two-class system. I don't celebrate it. I don't endorse it. I think those inequities are wrong. But what's wrong is the inequities. It's not that they're biological. Any inequity that leads to these kinds of different abilities to enjoy and pursue life ought to be redressed. So what the inequity criticism misses is that what's wrong is inequity, not biology.

And worse, those people who keep telling us that they care about it so much do nothing to suggest rectification of environmental, social and familial inequity. They have nothing to say. It's only if I put a chip in my head. I can attend Harvard all day, apparently, and come up with the $40,000 it takes to go there and they don't care. But if it's some kind of intervention that might be biochemical, or bioengineered, that, for some reason, is a different kind of inequity and they don't like that.

That is not treating like cases alike; that's an old principle of morals. And I think the argument falls down here completely. Inequity is the problem, not biological engineering.

Satisfaction guaranteed?

Well, what about this whole idea that it is wrong, that we will find ourselves unhappy, dispirited and dissatisfied if we have cheap victories? If we wind up using biological knowledge to engineer ourselves so that we can think more quickly in solving a problem, have more memory, figure out problems better than we could before because we've taken a drug? What if drugs out now like modafinil (Provigil) allow us to sleep less?

If we swallow a cup of coffee every morning and use that stimulant, should we all feel morally bad for a while? I mean, that's what the argument is. You're making a pharmacological intervention to get your attention going. And apparently, that's a cheap thrill that you don't really deserve. You should just wake your own damn self up and run around the block a few times rather than having these shortcuts.

Some people do think that the only way to get to the top of the mountain is to hike up there. I don't have a problem with that. If they like doing that, that's fine. Me, I like a helicopter. View's the same. I don't care. I get to the top. I get to see it. It's faster. I leave.

Now there are other things that I don't understand at all. Golf! Take a little ball. You've got a stick. You run around. You hit it in the hole. It's satisfying to you. I don't know. Whatever. I mean, I'm not saying we can't delude ourselves into feeling that certain pursuits that we work at and craft and hone (jogging, golf) give us enjoyment, that we get a sense of testing our limits, a test of pushing ourselves and that gives us pleasure.

But what I am saying is that's not all forms of pleasure. There are plenty of things that you and I are all happy about that we have nothing to do with, that we don't struggle, practise, earn, fight for or do anything to attain. They just happen and we say, 'Well, that's good fortune.' Part of that, obviously, is a lesson of Darwinism.

I'm tempted to say, 'Stuff happens.' But there's a lot of stuff out there, the genetic lottery, so that you might say, 'Boy, I'm glad I can sing. Or I'm glad that I have good pitch. Or I'm really lucky that I have good hand–eye coordination.' I'm describing a number of things I don't have and I'm envious of. But I don't begrudge them. That's just how it is. You're lucky to have those things, and so that's nice. And I'm happy. And you're happy. And we don't sit around saying it's only earned happiness is the authentic happiness. I don't buy that argument at all. I think it's a distortion, in fact, of what makes human beings satisfied.

Creatures in flux

So what remains in our march against the anti-meliorist here is this human nature argument. At bottom, the other ones, I think, collapse. They're not good arguments about why we shouldn't try to improve ourselves. Human nature is probably the last bastion of defence for the anti-meliorists.

We're a jumble of traits, behaviours, aptitudes, interests, capacities and volitions shaped by a set of historical accidents. The problem is that certain conservatives who would like to anchor the world keep trying to resurrect Platonic essentialism. In fact, the real worry for Darwinism isn't the scientific creationists, isn't the anti-evolutionists. They're just people who want religion introduced into people's lives.

The real threat is the anti-meliorists. Those who argue for a distinct essence, a kind of template of humanity that somehow is in there as a core that cannot be touched or changed or manipulated without loss of who we are – they are nervous conservatives who worry that the bearings will be lost if we admit that what we are is a jumbled set of mishmash traits evolved and designed to handle a random environment from the past that we don't have to care about any more.

The anti-meliorists don't tell us what human nature is. They posit a static notion of human nature which isn't consistent with evolution. They posit the view that what our nature is, whatever it is, is right, when we know that it's right only in relation to a set of environmental challenges that don't exist any more or that we're modifying all the time.

We are a creature or species, as all are, in a state of flux. The anti-meliorists are making the conceptual error, that the way we are is the way we should be. I'm submitting that what we know from evolution, from Darwin's day on, is that the way we are is an interesting accident. And it tells us certain things about what will make us function well, but it doesn't tell us anything about the way we should be or what we should become or how we should decide to change ourselves.

I find no in-principle arguments why we shouldn't try to improve ourselves at all. I don't find it persuasive that to say you want to be stronger, faster, smarter makes you vain. Try to improve yourself. From Ben Franklin on, there are both secular and religious thinkers who urge improvement on our species and our individual selves.

That's what agriculture is. That's what plumbing is. That's what clothes are. That's what transportation systems are. They are all attempts by us to transcend our nature. Do they make us less human? Or are they the one possible contender for what it means to be human? This idea that we want to try and press change, improve. Maybe that's it. If that's it, I'll accept that because I think that may be the only thing we can draw out of evolution.

If we limit ourselves, in the way that many anti-meliorists are suggesting we do now, then we will rob ourselves and our descendants of some of the most exciting opportunities that the biological revolution presents.

Professor Caplan is Director of the Center for Bioethics at the University of Pennsylvania. This essay is adapted from his Charles Darwin Memorial Lecture given on 17 February 2005 at The General Society of Mechanics and Tradesmen of the City of New York.

Notes

1 President's Council on Bioethics, *Beyond Therapy: Biotechnology and the pursuit of happiness* (Washington, DC, Oct 2003).
2 M Sandel, 'The case against perfection', *Atlantic Monthly*, Apr 2004.

3. Welcome to a world of exponential change

Nick Bostrom

For most of human history, the pace of technological development was so slow that a person might be born, live out a full human life and die without having perceived any appreciable change. In those times, worldly affairs appeared to have a cyclical nature. Tribes flourished and languished, bad rulers came and went, empires expanded and fell apart in seemingly never-ending loops of creation and destruction. To the extent that there was a direction or destination to all this striving, it was commonly thought to lie outside time altogether, in the realm of myth or supernatural intervention.

A present day observer, by contrast, expects to see significant technological change within a time span as short as a decade and much less in certain sectors. Yet although the external factors of the human condition have been profoundly transformed and continue to undergo rapid change, the internal factors – our basic biological capacities – have remained more or less constant throughout history. We still eat, sleep, defecate, fornicate, see, hear, feel, think and age in pretty much the same ways as the contemporaries of Sophocles did. But we may now be approaching a time when this will no longer be so.

The prospect of artificial intelligence

The annals of artificial intelligence are littered with broken promises. Half a century after the first electric computer, we still have nothing that even resembles an intelligent machine, if by 'intelligent' we mean possessing the kind of general-purpose smartness that we humans pride ourselves on. Maybe we will never manage to build real artificial intelligence. The problem could be too difficult for human brains ever to solve. Those who find the prospect of machines surpassing us in general intellectual abilities threatening may even hope that is the case.

However, neither the fact that machine intelligence would be scary nor the fact that some past predictions were wrong is good ground for concluding that artificial intelligence will never be created. Indeed, to assume that artificial intelligence is impossible or will take thousands of years to develop seems at least as unwarranted as to make the opposite assumption. At a minimum, we must acknowledge that any scenario about what the world will be like in 2050 that postulates the absence of human-level artificial intelligence is making a big assumption that could well turn out to be false.

It is therefore important to consider the alternative possibility: that intelligent machines will be built within 50 years. We can get a grasp of this issue by considering the three things that are needed for effective artificial intelligence. These are: hardware, software and input/output mechanisms.

The requisite input/output technology already exists. We have video cameras, speakers, robotic arms etc that provide a rich variety of ways for a computer to interact with its environment. So this part is trivial.

The hardware problem is more challenging. Speed rather than memory seems to be the limiting factor. We can make a guess at the computer hardware that will be needed by estimating the processing power of a human brain. We get somewhat different figures depending on what method we use and what degree of optimisation we assume, but typical estimates range from 100 million MIPS to 100 billion MIPS (1 MIPS = one million instructions per second). A high-range PC today has a few thousand MIPS. The most powerful supercomputer to date performs at 260 million MIPS. This means that we will soon be within striking distance of meeting the hardware requirements for human-level artificial intelligence. In retrospect, it is easy to see why the early artificial intelligence efforts in the 1960s and 1970s could not possibly have succeeded – the hardware available then was pitifully inadequate. It is no wonder that human-level intelligence was not attained using a less-than-cockroach level of processing power.

Turning our gaze forward, we can predict with a high degree of confidence that hardware matching that of the human brain will be available in the foreseeable future. We can extrapolate using Moore's Law, which describes the historical growth rate of computer speed. (Strictly speaking, Moore's Law as originally formulated was about the density of transistors on a computer chip, but this has been closely correlated with processing power.) For the past half century, computing power has doubled every 18 months to two years. Moore's Law is really not a law at all, but merely an observed regularity. In principle, it could stop holding true at any point in time.

Nevertheless, the trend it depicts has been going strong for an extended period of time and it has survived several transitions in the underlying technology (from relays to vacuum tubes, to transistors, to integrated circuits, to very large integrated circuits). Chip manufacturers rely on it when they plan their forthcoming product lines. It is therefore reasonable to suppose that it may continue to hold for some time. Using a conservative doubling time of two years, Moore's law predicts that the upper-end estimate of the human brain's processing power will be reached before 2020. Since this represents the performance of the best supercomputer in the world, one may add a few years to account for the delay that may occur before that level of computing power becomes available for doing experimental work in artificial intelligence. The exact numbers don't matter much here. The point is that human-level computing power probably has not been reached yet, but almost certainly will be attained well before 2050.

This leaves the software problem. It is harder to analyse in a rigorous way how long it will take to solve that problem. (Of course, this holds equally for those who feel confident that artificial intelligence will remain unobtainable for an extremely long time – in the absence of evidence, we should not rule out either alternative.) Here we will approach the issue by outlining just one approach to creating the software, and presenting some general plausibility arguments for how it could work.

We know that the software problem can be solved in principle. After all, humans have achieved human-level intelligence, so it is evidently possible. One way to build the requisite software is to figure out how the human brain works, and copy nature's solution.

It is only relatively recently that we have begun to understand the computational mechanisms of biological brains. Computational neuroscience is about 15 years old as an active research discipline. In this short time, substantial progress has been made. We are beginning to understand early sensory processing. There are reasonably good computational models of the primary visual cortex, and we are working our way up to the higher stages of visual cognition. We are uncovering what the basic learning algorithms are that govern how the strengths of synapses are modified by experience. The general architecture of our neuronal networks is being mapped out as we learn more about the interconnectivity between neurones and how different cortical areas project onto one another. While we are still far from understanding higher-level thinking, we are beginning to figure out how the individual components work and how they are hooked up.

Assuming continuing rapid progress in neuroscience, we can envision learning enough about the lower-level processes and the overall architecture to begin to implement the same paradigms in computer simulations. Today, such simulations are limited to relatively small assemblies of neurones. There is a silicon retina and a silicon cochlea that do the same things as their biological counterparts. IBM's 'Blue Brain Project' aims to create an accurate software replica of a neocortical column by 2008. Simulating a whole brain will of course require enormous computing power; but as we saw, that capacity will be available within a couple of decades.

The product of this biology-inspired method will not be an explicitly coded mature artificial intelligence. (That is what the so-called classical school of artificial intelligence tried unsuccessfully to do.) Rather, it will be a system that has the same ability as a toddler to learn from experience and to be educated. The system will need to be taught in order to attain the abilities of adult humans. But there is no reason why the computational algorithms that our biological brains use would not work equally well when implemented in silicon hardware.

The promise of nanotechnology

In 2005, Europe, the US and Japan spent approximately one billion US dollars each on nanotechnology in public funding, a tenfold increase since 1997.[1] 'Nanotechnology' has become a buzzword. Putting it into a grant application can greatly increase its chance of being funded – as Oxford Professor George Smith jokes, 'nano is from the Greek verb meaning "to attract research funding"'.

The word was coined by Dr Erik Drexler and popularised in his 1986 book *Engines of Creation*.[2] Drexler published detailed technical analyses arguing for the feasibility of building molecular machines to atomic precision.[3] Such machine-phase nanotechnology, in its mature form, will give humanity unprecedented control of the structure of matter. In many respects, it will transform manufacturing into a software problem. In Drexler's vision, nanotech construction devices would build objects one molecule at a time, and billions of such devices working in parallel would be able to construct atomically almost-perfect objects of arbitrary size. Applications would include:

- extremely fast computers
- lighter, stronger materials (a strong enabling factor for space technology)
- clean, efficient manufacturing processes of most products
- cheap solar energy production, and the ability to actively scrape excessive CO_2 out of the atmosphere
- desktop manufacturing devices with near-universal capabilities

- tiny medical robots that could enter individual cells and perform molecular-level repair, eliminating most disease and ageing, and making it possible to upload human minds to computers (this would be a second possible route to human-level artificial intelligence).

Drexler noted that a technology this powerful could also be used with devastating results for destructive ends. He worried especially about dangerous arms races, new weapons of mass destruction that could be used by terrorists and rogue states, and mind-control technologies that could be used by bad governments to oppress their populations. Nevertheless, Drexler argued, attempts to prevent the development of nanotechnology would necessarily fail and would on balance increase the dangers.

Although Drexler helped create the enthusiasm for the field of nanotechnology that has resulted in the recent funding boom, he has subsequently been sidelined by the mainstream community of nanoscientists because his vision runs too far ahead of the experimental work that is currently being done in labs and the applications that are immediately on the horizon. Another reason for Drexler's marginalisation is the fear felt by some nanotechnologists that the future dangers to which he drew attention could fan public opposition to nanotechnology, resulting in a loss of funding. One Nobel laureate, Richard Smalley, declared – without offering any technical argument – that Drexler's vision was physically impossible and went on to accuse Drexler of 'scaring our children'.[4]

More recently, there are some signs that the Drexlerian vision of nanotechnology might be poised for a comeback, thanks partly to rapid scientific progress in the field and new computer modelling studies that seem to support the feasibility of molecular machine systems. Policy-makers are already concerned with the need to examine the far-reaching ethical and social implications that nanotechnology will have once it is fully developed. For example, a nanotechnology bill signed into law by President Bush in late 2003 requires that the programme ensure

> *that ethical, legal, environmental, and other appropriate societal concerns, including the potential use of nanotechnology in enhancing human intelligence and in developing artificial intelligence which exceeds human capacity, are considered during the development of nanotechnology.*[5]

Some 3 per cent of the budget of the Human Genome Project was devoted to studying the ethical, legal and social issues (ELSI) around the availability of genetic information. It looks like nanotechnology is set to continue this new trend of including the social science and the humanities in major technological research programmes. Such anticipatory ELSI research is a new phenomenon, and its long-term effects remain to be seen.

Convergence and the singularity

The concept of 'converging technologies' stems from a 2002 report sponsored by the US National Science Foundation (NSF), and edited by Mihail Roco and William Bainbridge:

> *In the early decades of the twenty-first century, concentrated efforts can unify science based on the unity of nature, thereby advancing the combination of nanotechnology, biotechnology, information technology, and new technologies based in cognitive science. With proper attention to ethical issues and societal needs, converging technologies could achieve a tremendous improvement in human abilities, societal outcomes, the nation's productivity, and the quality of life.*[6]

The phrase 'converging technologies' refers to the synergistic combination of four major provinces of science and technology, known in short as 'NBIC'. These are (a) nanoscience and nanotechnology; (b) biotechnology and biomedicine, including genetic engineering; (c) information technology, including advanced computing and communications; and (d) cognitive science, including cognitive neuroscience. The idea is that as these four areas develop they will join to create a more integrated approach to science and technology, where, for instance, the boundary between biotechnology and nanotechnology dissolves. The NSF report describes how dramatic new capabilities would result and could be used to enhance human capacities.

It has been said that most people overestimate how much technological progress there will be in the short term and underestimate how much there will be in the long term. There is usually a long lag time between proof-of-concept in some laboratory and the time when actual products begin to have a significant impact in the market. Many a seemingly good idea never pans out. Hot technological fields usually yield a lot of hype.

The world economy is doubling every 15 years. Particular technological areas exhibit faster growth. Ray Kurzweil, the American inventor and technology forecaster, has documented many technological areas, including computing, data storage, gene sequencing, brain mapping and others, where progress is currently occurring at a rapid exponential pace. Exponential growth starts slow and then becomes very fast. Here is a classic problem that illustrates this:

> *The water lilies in a pond double every day. It takes two weeks before the lilies cover the whole pond. How long did it take before they covered half of the pond?*

The answer, of course, is that the exponentially growing lily population covered half the pond on day 13, one day before it doubled again to cover the whole pond. Kurzweil argues that we intuitively tend to think of progress as linear while in reality it is exponential, and that many people will be surprised to find how rapidly things develop over the coming decades. Kurzweil believes that we will not experience 100 years of progress in the twenty-first century – it will be more like 20,000 years of progress (at today's rate).[7] This is because our ability to invent new things is itself improving, through advances in scientific instrumentation, methodology and computing.

The 'singularity' is a hypothetical point in the future where the rate of technological progress becomes so rapid that the world is radically transformed virtually overnight. The only plausible scenario in which such a singularity could occur is through the development of machine intelligence. One might imagine that machines will at some point come to significantly surpass biological human beings in general intelligence, and that these machines will be able to apply their intelligence to rapidly improve themselves so that within short order they become superintelligent. Superintelligent machines would then be able to rapidly advance all other fields of science and technology. Among the many other things that would become possible is the uploading of human minds into computers, and dramatic modification or enhancement of the biological capacities of human beings that remain organic.

It is of course an open question whether a singularity will ever occur. It is possible that there will never be a point where progress becomes as rapid as the singularity hypothesis postulates. Even if there were to be a singularity at some point, it is very difficult to predict how long it would take to get there, although some have argued that it is more likely than not that we will have superintelligent machines before the middle of the twenty-first century.[8]

What is a policy-maker to do in light of all these possibilities? A first priority is to abandon the unquestioning assumption that human nature and the human condition will remain fundamentally unchanged throughout the current century. A second is to develop better techniques for long-range planning and horizon-scanning. Such techniques are already used in some policy decisions, for example, in arguments about the importance of reducing global warming. Yet once we consider the bigger picture, we may feel that the risks of global warming are dwarfed by other risks that our technological advances will create over the coming several decades.[9] Perhaps we ought to spend a fraction of the money and effort currently devoted to the problem of climate change to thinking about these other risks too.

And in addition to risks, there are also immense opportunities. Again, consideration of the big picture can help us spot opportunities for saving lives and improving the quality of life that might otherwise go unnoticed. A massive increase in funding for research to better understand the basic biology of ageing could pay off handsomely if it leads to treatments to intervene in the negative aspects of senescence, allowing men and women to stay healthy and economically productive much longer than is currently possible.

That there will be change is certain, but what the change will be depends in some measure on human choice. In this century we may choose to use our technological ingenuity to unlock our potential in ways that were unimaginable in the past.

Nick Bostrom is Director of the Future of Humanity Institute at the University of Oxford and Chair and co-founder of the World Transhumanist Association. The section of this article on artificial intelligence is based on 'When machines outsmart humans' by Nick Bostrom in Futures *35, no 7 (2003).*

Notes

1 M Roco and WS Bainbridge (eds), *Nanotechnology: Societal implications maximizing benefits for humanity*. Report of the National Nanotechnology Initiative Workshop (Arlington, VA, 2003).
2 KE Drexler, *Engines of Creation: The coming era of nanotechnology* (London: Fourth Estate, 1986).
3 KE Drexler, *Nanosystems: Molecular machinery, manufacturing, and computation* (New York: John Wiley & Sons, Inc., 1992).
4 E Drexler and R Smalley (2003). 'Nanotechnology: Drexler and Smalley make the case for and against "molecular assemblers"', *Chemical & Engineering News* 81, no 48 (2003).
5. '21st Century Nanotechnology Research and Development Act' (passed 3 Dec 2003) (1, section 2.B.10); see: www.nano.gov/html/news/PresSignsNanoBill.htm (accessed 17 Jan 06).
6 MC Roco and WS Bainbridge (eds), *Converging Technologies for Improving Human Performance* (Arlington, VA: National Science Foundation/Department of Commerce-sponsored report, 2002).
7 R Kurzweil, 'The law of accelerating returns', KurzweilAI.net, 2001. Available at: www.kurzweilai.net/articles/art0134.html?printable=1 (accessed 7 Jan 2006).

8 See N Bostrom, 'How long before superintelligence?', *International Journal of Futures Studies* 2 (1998), and H Moravec, *Robot: Mere machine to transcendent mind* (New York: Oxford University Press, 1999).
9 N Bostrom, 'Existential risks: analyzing human extinction scenarios and related hazards', *Journal of Evolution and Technology* 9 (2002).

4. The man who wants to live forever

Paul Miller and James Wilsdon

It's a bright October morning when Aubrey de Grey meets us at Cambridge station. The leaves are falling in the wind and mounting up in small piles on the station forecourt. de Grey is standing by his battered red racing bike, wearing scruffy white trainers, a fleece sweatshirt and jeans. His beard makes him easy to spot – a swirling mass of brown, red and grey hair reaching well below his chest.

As we walk through the station car park he asks us about Demos – what it's like, how it's funded, what else we're working on. He didn't get the email we'd sent earlier that morning as he tends to get up late and work through the night. 'At least it means that I'm awake at the right time when I go to conferences in America,' he says. He tells us that in 2005 he gave 33 invited talks abroad.

After a short walk, he locks his bike up outside a pub and marches inside. The barman gives a nod of recognition. 'I've done loads of interviews here,' he tells us, 'usually at this table.' He jokes with journalists that he will buy them a pint if they ask him a question he hasn't heard before.

What he has said at this table in this pub (as well as in journals and at conferences around the world) has caused no end of controversy. de Grey's claim is that radical increases in human life expectancy will be possible within the next 20 to 30 years. 'As medicine becomes more powerful', he says, 'we will inevitably be able to address ageing just as effectively as we address many diseases today. . . . I think the first person to live to 1000 might be 60 already.'

The basis for such a confident prediction is a project that he calls Strategies for Engineered Negligible Senescence. It makes SENS to put it into practice, he jokes. The SENS project, which he directs, has identified seven causes of ageing – seven types of molecular or cellular damage – each of which 'is potentially fixable by technology that already exists or is in active development'.

de Grey argues that society is caught in a 'pro-ageing trance', which leads most of us to defend 'the indefinite perpetuation of what it is in fact humanity's primary duty to eliminate as soon as possible'. His forthright views, and the endearing zeal with which he expresses them, have attracted increasing amounts of attention, not only in scientific journals but also in the mainstream media. In the past year, he has been interviewed by several UK broadsheets, and he recently featured in a *60 Minutes* special on longevity on the US network CBS.[1] In a profile of de Grey for the magazine *Technology Review*, Sherwin Nuland, Clinical Professor of Surgery at Yale, concluded that 'his stature has become such that he is a factor to be dealt with in any serious discussion of the topic'.[2]

A prophet without honour?

Within the scientific community, de Grey is regarded with a mix of interest and scepticism. Is he a pioneer or a crank? Naïve or prophetic in his claims that we will soon be able to live for hundreds of years? Nuland's now infamous profile in *Technology Review* implied that there were major flaws in de Grey's scientific theories, but seemed more concerned that he might pose a threat to society. Noting that 'the most likeable of eccentrics are sometimes the most dangerous', Nuland concluded that

> *his clarion call to action is the message neither of a madman nor a bad man, but of a brilliant, beneficent man of goodwill, who wants only for civilisation to fulfil the highest hopes he has for its future. It is a good thing that his grand design will almost*

> *certainly not succeed. Were it otherwise, he would surely destroy us in attempting to preserve us.*[3]

In the firestorm that followed the *Technology Review* piece, things got personal. One editorial comment – likening de Grey to a troll – still reverberates in internet discussions.

'Do you care what people say about you?' we ask. 'Yes, deeply,' is his instant reply. 'I take it very seriously.' Yet de Grey says that he's moved beyond the stages of being ignored or laughed at and is now being actively opposed. He seems quite relaxed about this progression.

We talk about Eric Drexler, the US scientist who coined the term nanotechnology, but is now shunned by the field that he helped to create. Back in 1995, the BBC screened an edition of the *Horizon* programme where a buoyant Drexler comes across as an optimistic proponent of nanotechnology and the age of molecular manufacturing. Roll on ten years and he has become a bitter figure, marginalised by the mainstream nanoscience community, and a veteran of long-running battles with US science funders who have refused to back his vision.

Is de Grey worried that he may suffer the same fate as Drexler within his own field of biogerontology? 'I have a massive advantage,' he says, 'I'm integrated into the mainstream of gerontology. I rose very rapidly to become considered an intellectual equal of the leading people in the field. I go to all the international events and everyone knows I'm not a fool. In person, there's always a degree of cordiality and respect, even though there's ostensibly a much more caustic debate about my work in print.'

He admits there are risks in being labelled a 'transhumanist'. 'The sort of people who have been at the forefront of talking about life extension have also been talking about things that people are much more comfortable dismissing – like nanotechnology and cryonics. There's been a feeling that they're not quite one of us.' He says he started out deeply sceptical about transhumanism as an idea but has warmed to it over the years. He recently signed up with the Arizona-based company Alcor to be cryonically frozen – presumably as a backup if the technology for life extension doesn't come along quite as soon as he hopes.

We ask him whether he sees himself as a campaigner. 'Crusader', he says with a smile. 'I just want to save lives. I see no difference between preventing someone's death through medicine and preventing death through defeating ageing. It's just not a distinction.'

He certainly doesn't fit the mould of a traditional science communicator. He couldn't be further from the model of Robert Winston or Simon Singh, explaining science to the masses. It's difficult, for example, to imagine him wearing a dinner jacket and delivering the Royal Institution Christmas lectures. He relishes debating and talking about the social and political implications of his science. In many ways he's an advocate of what Demos calls 'upstream public engagement', of sparking a wider debate about the potential implications of scientific and technological change, before those changes are locked into immovable trajectories.

And he's very good at answering questions. He speaks quickly but in perfectly formed paragraphs, uncluttered by ums and errs.

Who wants to live forever?

During the 1994 Miss America competition, the host asked Miss Alabama: 'If you could live forever, would you want to, and why?' Miss Alabama answered, 'I would not live forever, because we should not live forever, because if we were supposed to live forever, then we would live forever, but we cannot live forever, which is why I would not live forever.'[4]

The question that de Grey has to field most often is 'do people really want to live much longer lives?' Miss Alabama's answer reflects the confused, but instinctive response that many of us have to the idea of having radically extended lifespans. Bill McKibben – a more articulate 'deathist', as the transhumanists deride their critics – puts it simply, 'I like this planet, I like this body with all its limitations, up to and including the fact that it's going to die.'[5]

de Grey isn't convinced. As far as he's concerned, this is another symptom of the pro-ageing trance: 'It's a coping strategy. Ageing isn't much fun, getting decrepit and senile. You have to find some way of putting it out of your mind. But we're talking about the extension of healthy life, not just extending old age. Psychologically it's terribly difficult for people to take on board that this is something worth fighting for.'

He thinks people will eventually come round to his way of thinking, arguing that the media's fascination with his theories tells you something about the pent-up demand for longer, youthful lives. And he argues that there are similarities to the way people dress to impress, use makeup, or resort to cosmetic surgery to make themselves attractive. As he puts it, 'Looking younger when you're older is no different to looking prettier when you're young.'

Unsurprisingly, some have expressed ethical concerns about what de Grey is proposing. 'People say "aren't you playing God by using technology to make people live longer?" The obvious answer is that, if that's the case, so was inventing the wheel. We are people who have the ability to improve our world if there's anything we don't like about it. God made us that way.' de Grey plays down the level of resistance he's faced from religious groups in particular: 'We're not talking about immortality here. If God still wants you to die you will, whether that's because of ageing or because you get hit by a truck. It's all the same to God. I've never – not once – had someone religious challenge that.'

The secret to longer life

Each year, de Grey tells us over another beer, journalists traipse to the house of the oldest woman in the world on her birthday to ask her the secret to her long life. Each year, she comes up with a different answer. de Grey's favourite is that it was because she gave up smoking when she was 111.

For de Grey, the secret to longer life lies in the research lab. His predictions centre on work that is now getting under way on mice. Once the landmark point of 'Robust Mouse Rejuvenation' is reached, politicians and the public will no longer be able to ignore his claims. This will be achieved when mice that would have a normal lifespan of three years can be given therapies two-thirds of the way through their life that allow them to live until they are five. Funding of around £50 million a year for the next ten years is what's required to get us there, and de Grey expects it would take a further 15 years for scientists to be able to transfer the therapies successfully from mice to humans.

Once mice have shown what could be achieved, 'it's impossible to imagine public funding not coming. There's no way that government will be able to walk away from it. It would make them unelectable.' de Grey likens the task of developing technologies for life extension to the Wright brothers' pioneering flying machines or the Apollo programme to send a man to the moon: 'There's a fundamental difference between the creativity required to do science and the creativity involved in developing technology. You take the problems apart, divide them into manageable chunks and solve them.'

Where does he think such funding might come from? 'Public funding tends to be low risk, low gain. And technology funding from venture capital is too short term. What we need is funding that is ambitious and long term. That tends to come when national pride is at stake or when seriously rich people think it's cool.'

It's the latter – an elite band of Silicon Valley millionaires and wealthy philanthropists – who are funding much of the research into life extension at the moment. We ask him whether people are right to be suspicious about the motivations of wealthy funders? Shouldn't there be more democratic oversight of such research? He shrugs his shoulders: 'What's the point in being wealthy if you can't make a difference? I'm not worried if it only comes from that kind of source in the beginning. When we can show that it will work for mice, public funding will come.'

At the moment, mainstream science funding eludes de Grey. But he's not too worried about it. 'Because of my position as a theoretician, I can piss anybody off and not worry about the repercussions for my next grant application.' Politics too remains a foreign world to him. 'Politicians work with the art of the possible and they are still being told by people they trust that Aubrey de Grey is a complete fruitcake.'

This is why he spends his time talking to journalists, even though, as he says, 'it's exhausting'. de Grey's aim is to 'lower the activation energy', so that the move from fruitcake to mainstream for his brand of biogerontology happens more quickly. 'I'm doing what comes naturally to me; just playing to my strengths. I think I'm good at enthusing people and I think I'm a good scientist. I'm not a politician or a social scientist and I'm no good at not pissing people off.'

But though he admits there aren't any votes in it yet, de Grey is convinced that politicians and policy-makers should think about life extension now rather than later. 'Most policy-makers get interested in therapies when they're at the human trial stage. That's wrong. They need to think about these things when we're at the mice stage, if not before. Life extension could go from zero to infinity faster than the web. It will have a massive impact on the way that people live and plan their lives. Just think about how tricky it's going to be to retain people in vital but risky jobs – like the fire service or the army. Everybody's going to be doing their best to live long enough to live forever.'

A public conversation

We shake hands outside the pub and Aubrey unlocks his bike, spins it around and heads off through Cambridge's narrow streets. Sitting on the train back to London, we admit that we are more impressed than we expected to be. de Grey's media persona, reflected in the growing canon of articles about him, fails to do justice to the subtleties of his position, and the strategic flair with which he is influencing a debate that may, just may, turn out to be one of the defining issues of our time. He is acutely aware of what he's doing. And whether scientists agree with him or not, there's no doubt that he is playing an important role. As biogerontologist Jay Olshansky says: 'I am a big fan of Aubrey. We need him. I disagree with some of his conclusions, but in science that's OK. That's what advances the field.'[6]

Science isn't a clean, logical endeavour pursued by individuals who interact only through peer-reviewed journals. It's a messy mixture of experimentation, argument and debate. And when it meets politics it becomes messier still. It is every scientist's responsibility to shape and be shaped by what society wants from science, to listen to the public and to take their concerns seriously. Whatever one may feel about his theories, this is something that Aubrey de Grey is doing in a quite unique and valuable way.

Notes

1 CBS, 'The quest for immortality', *60 Minutes*, 1 Jan 2006.
2 S Nuland, 'Do you want to live forever?', *Technology Review*, Feb 2005.
3 Ibid.
4 Cited in K Philipkoski, 'Who wants to live forever?', *Wired Magazine*, Nov 2002.
5 Bill McKibben, *Enough: Genetic engineering and the end of human nature* (London: Bloomsbury, 2003).
6 S Nuland, 'Do you want to live forever?'

5. The transhumanists as tribe

Greg Klerkx

Just a few years into the new century, Russian biologist Elie Metchnikoff believed he had discovered the key to achieving immortality. And, more excitingly, Metchnikoff was convinced that by doing so he was on the verge of ushering in a new phase of human evolution. Bypassing *elixir vitae* and the fountain of youth, Metchnikoff literally went with his gut. What stood between mortality and potential immortality, Metchnikoff claimed, was the large intestine, which he viewed as one of evolution's more dangerous leftovers: a cesspool of waste and, critically, the human body's primary breeding ground for bacteria. To Metchnikoff, bacteria were the real enemy. Remove them, and you remove one of the chief causes of natural death in humans.

Metchnikoff's idea quickly gained traction both in the scientific community and among the public. He gave speeches; he wrote books. And he performed experiments – on humans – to test his theory. He surgically removed the bowels of several ill patients and claimed they were the better for it, even though some promptly died.[1]

If Metchnikoff's methods seem outdated and extreme, it is probably because the 'new' century in which his life extension theory took hold was the twentieth, not the twenty-first. But far from being a well-meaning crank, Metchnikoff was among the most prominent scientists working at the dawn of biology's modern golden age. He was an associate of Louis Pasteur (and eventually a director of the

Pasteur Institute) and he shared with Pasteur an obsession with microbes and their role in disease. Metchnikoff made his own indelible contribution to biology by identifying phagocytes, or white blood cells: the body's first line of defence against infection. For this achievement, Metchnikoff was awarded the 1908 Nobel Prize for Physiology/Medicine.

Metchnikoff didn't shy away from suggesting that science had brought humankind to the brink of a new and remarkable era. 'The human condition as it exists today, being the result of a long evolution and containing a large animal element, cannot furnish the basis of rational mortality,' he wrote in his 1910 book, *The Prolongation of Life*. 'The conception which has come down from antiquity to modern times . . . is no longer appropriate to mankind.'[2]

In most of the ways that count, Metchnikoff was the first modern transhumanist. At the least, he was the first modern populariser of a very old aspiration: to use technology and, later, science to transcend what nature has endowed us. He also neatly framed transhumanism as a temporary state between old and new: between the incremental progress of natural development and a future in which humans took every aspect of their destiny, including their biology, firmly into hand. But even this idea, though seemingly rooted in modern bioscience, has very ancient antecedents: Icarus's wings were, if nothing else, an early expression of a primitive transhumanist yearning.

Modern day Metchinikoffs

These days, transhumanists take many forms: from nanotech enthusiasts who envision armies of microscopic robots inside our bodies, forever detecting and destroying disease, to head-freezing cryonicists who believe that science will one day revive the dead. But all share a basic belief that would undoubtedly resonate with Metchnikoff: that as technology and medicine advance, humans will live significantly longer and healthier lives while realising greater intellectual and social achievements. As a result, there will be a profound change in what it means to be human.

The term 'transhumanism' had no real purchase on popular

culture in Metchnikoff's day, although references to it can be found as early as 1312, when in *The Divine Comedy* Dante Alighieri used *transumanar* to describe what happens to someone who experiences a holy vision.[3] But our modern sense of the word is more clearly associated with the revolution in biological science of which Metchnikoff was an early leader. Even in Metchnikoff's day, the headiness of this revolution, which seemed to match fantastic claims with amazing achievements, infected both the academy and the masses. Reflecting on Metchnikoff and his writings, in 1903 the *Times* confidently stated that, 'We should live till 140 years of age. A man who expires at 70 or 80 is the victim of accident cut off in the flower of his days, and he unconsciously resents being deprived of the 50 years or so which Nature owes him.'[4]

Certainly, Metchnikoff's position as one of the world's pre-eminent biologists helped give his ideas currency, but today's transhumanists do not lack their eminent 'mainstream' representatives. Ray Kurzweil, one of the most vocal promoters of transhumanism, is an accomplished technologist who has been awarded prize after prize for inventions as important as the flatbed scanner and machines that help the blind use computers. Another eminent transhumanist is Marvin Minsky, founder of the MIT Media Lab and a leading light in the development of artificial machine intelligence. Kurzweil in particular has been successful, with books and talks, in painting a convincing picture of a near-term world in which humans will be repaired, enhanced and advanced by bioscience to such an extent that our children, or theirs, will effectively be immortal.

Kurzweil, Minsky and their peers are at the forefront of what might be called transhumanism's third wave. Like Metchnikoff, they are sounding a clarion call that radically improved and longer-lived humans are imminent, and they are basing such claims on optimistic extrapolations from relatively new science and technology. Whereas Metchnikoff was excited by microbes, Kurzweil and company are enthusiastic about the possibilities deriving from rapid advances in computer and materials technology and the decoding, in 2000, of the human genome.

But the new transhumanists differ from Metchnikoff in a critical aspect. They are convinced that transhumanism is not a surprising byproduct of modern science; they believe it is an evolutionary inevitability and, critically, the only way in which humankind can be saved from its worst impulses. In this, what fuses the first and third waves of the transhumanist movement is the second wave, which gave rise to the modern definition and common use of the term itself.

Second wave optimism

At the head of this second wave of transhumanism was a former Olympic athlete turned novelist and futurologist who began life with the name Fereidoun M Esfandiary but ended it, in 2000, with a far more ethereal moniker: FM-2030. The son of an Iranian diplomat, Esfandiary had lived in 17 countries by the time he was 11 years old, and while he would spend most of his life in the United States, his early nomadic existence clearly defined him and the philosophy he would bring to the transhumanist *oeuvre*. As a reviewer of his first novel, the best-selling *Day of Sacrifice* (1959), wrote, 'Esfandiary is an optimist. He has hope, because he has a deep faith in man. He is convinced that technological progress, the contact of cultures, etc . . . will free man from his present miseries. Given time, man will even deliver himself from his supreme tragedy – death. Man can be made perfect.'[5]

Optimism was one of Esfandiary's critical contributions to the progress of transhumanism. The other was inevitability. Esfandiary was convinced that longer-lived humans were a necessary byproduct of the wave of sci-tech breakthroughs that had rocked the twentieth century. Conveniently, he dismissed the era's darker technological products, like nuclear weapons, as aberrations of human progress. To Esfandiary, radically extended life, not to say immortality, was the essential next step in humanity's escape from the randomness of natural evolution to a new place where it would assume true control of its destiny.

Esfandiary was not the first to espouse this viewpoint. In his 1957 essay 'Transhumanism', biologist Julian Huxley used the term to describe a future point when humankind would find itself 'on the threshold of a new kind of existence, as different from ours as ours is from that of Peking man.'[6] But Esfandiary took Huxley's essentially evolutionary cast of transhumanism and moved it a step nearer to the *zeitgeist*: the tipping point from human to transhuman didn't exist in the hazy future, Esfandiary insisted. It was already happening. 'Today when we speak of immortality and of going to another world we no longer mean these in a theological or metaphysical sense,' Esfandiary wrote in his 1973 book, *Up-Wingers*, which largely set the tone for all transhumanism to come, 'We now need new conceptual frameworks and new visions to guide us as we venture into uncharted spheres which are potentially full of hope.'[7]

At the time, few scoffed at Esfandiary's radical claims for an imminent transhuman awakening. The years preceding *Up-Wingers* had seen the introduction of the birth control pill and humans landing on the moon; the term 'Up-Winger' was a specific reference to spaceflight, which Esfandiary saw as a harbinger of the transhumanist revolution. Political establishments seemed bereft of answers to the woes of the planet; in the future, science would lead the way, and society would follow. The use of the word 'up' was Esfandiary's deliberate attempt to redefine human ambitions in the context of 'the right–left establishment'.

The years following the 'Up-Winger Manifesto', in which Esfandiary published best-selling books with titles like *Telespheres* (1977), *Optimism One: The emerging radicalism* (1970) and *Are You a Transhuman?* (1989), would see the first artificial heart transplant, the first use of genetic engineering, the popular emergence of the internet, exponential advances in computing technology, and the embryonic demonstrations of artificial machine intelligence. Thus, Esfandiary's brand of transhumanism advocated a deliberate and aggressive acceleration of the pace at which human science and technology took positive control of the world: controlling weather cycles, manipulating human biology and colonising planets were just the beginning.

Only a few years before *Up-Wingers* was published, the first human was cryonically 'suspended', an action seen by many as the first modern act of applied transhumanism. Esfandiary himself, after dying of pancreatic cancer in 2000 – and thus falling short of living until 2030, as his moniker ambitiously proclaimed – had himself placed in suspension at the Alcor Life Extension Foundation, Arizona. He remains there to this day in the hope that, some day, science and technology will become sufficiently advanced to bring him back to life.

Most scientists don't believe that Esfandiary or any of his fellow 'cryonauts' will ever be anything more than expensively frozen flesh. Like the cryonauts themselves, by the late 1970s, Esfandiary's brand of optimistic transhumanism was largely spent as a cultural force. Its fading was gradual but, in many ways, predictable. However hopeful Esfandiary himself might have been about the human condition, the transhumanist movement he created was philosophically yoked to other utopian movements of the late 1960s and early 1970s that were effectively hoisted on their own high-tech petards. The Space Age didn't deliver the population relief, societal unity or new energy sources that it once promised, via massive rotating colonies and mining operations on the moon. Neither did the Whole Earth movement, which was largely sparked by the astounding pictures of the Earth from space, significantly slow human development or our rapaciousness for environmental resources.

More than human

Throughout the 1980s and 1990s transhumanism, like dreams of colonising the stars and achieving Gaian connectedness, was largely the province of fringe organisations. Most prominent and influential among these was the Extropy Institute, founded in 1988 by Bristol native Max More, who positioned transhumanism as something actively pursued by increasing numbers of people. 'Transhumanism reaches beyond the sphere of humanism in its goal to improve the human condition,' More wrote. 'We seek to improve ourselves and the species of "human".'[8] (As for 'extropy', it's an optimistically loaded neologism – an intended antonym to entropy – that neatly reflects More's determinist view of transhumanism.)

From the start, More saw the potential of information technology to spread his gospel of transhumanism. He launched the Extropians List, an ongoing discussion of transhumanist issues, in 1991 – the same year that Sir Tim Berners-Lee established the World Wide Web. Since then, the institute's site has expanded to become one of the more comprehensive sources of information on all things transhuman. This also helped to establish transhumanism as an idea for the twenty-first century, in concert with the explosion of the internet and its quasi-utopian trappings. It is no surprise that many of the most enthusiastic modern transhumanists are also internet pioneers. Some, like Oracle Software founder Larry Ellison, are substantial funders of transhumanist research projects.

Indeed, it seems likely that first true 'transhuman' will be someone like Larry Ellison, who combines the ambition, willpower and wealth to achieve a new lease on life. In this respect, modern transhumanism is less utopian than its previous iterations, and more reflective of an atomising society in which only the strong (for which, read: very rich) survive. By contrast, Esfandiary and the other leading lights of transhumanism's second wave were, essentially, New Age socialists. Enhancement leading to virtual immortality was to be for all, to the betterment of the species. There would be no haves and have-nots.

Transhumanism's third wave didn't begin in earnest until 2000 with the decoding of the human genome. Already, it combines a dizzying array of scientific disciplines and research spanning the globe. It also encompasses political and cultural faultlines, with issues ranging from the availability of AIDS drugs in Africa to the opposition, among Christian conservatives, to stem cell research in the US. Given the complexity of modern transhumanism, it is perhaps no surprise that the third wave is exemplified by someone who feels equally at ease working within the methods and frameworks of bioscience, engineering and even philosophy. Like Metchnikoff, Cambridge genetics researcher Aubrey de Grey (profiled earlier in this volume) is a scientist, holding a PhD in biology. Like Kurzweil and Marvin, de Grey is also a technologist, a software engineer who ran a high-tech start-up company in the 1990s. And like

More and Esfandiary, de Grey does not shrink from making radical transhumanist claims that enrage as often as they attract: more than once, he has claimed that science and technology are close to achieving breakthroughs that will allow humans to live for 1000 years.

But even if we can use science and technology to extend our life spans and natural abilities, the big unanswered question is do we really want to? Third wave transhumanists cannot see many downsides to these developments, despite ample evidence that the products of modern science have been used at least as often for harm as for good. At the end of the day, they insist we will become transhuman simply because it is our destiny. 'We didn't stay on the ground, we didn't stay on the planet, we're not staying within the limits of our biology,' says Kurzweil. 'We're a species that instinctively seeks to go beyond our limitations.'[9]

Greg Klerkx is a science writer and the author of Lost in Space: The fall of NASA and the dream of a new space age *(New York: Pantheon Books, 2004). Formerly with the SETI Institute, he is now at work on a play based on the life of rocket pioneer Wernher von Braun.*

Notes

1 Details about Elie Metchnikoff's research are from MR Rose, *The Long Tomorrow: How advances in evolutionary biology can help us postpone aging* (London/New York: Oxford University Press, 2005).
2 E Metchnikoff, *The Prolongation of Life* (London: W Heinemann, 1910).
3 From the *Paradiso*, Canto I, verses 64–72.
4 Preface to E Metchnikoff, *The Nature of Man: Studies in optimistic philosophy* (London: W Heinemann, 1903).
5 From www.fm2030.com (accessed 4 Jan 2006), a website dedicated to the life and work of FM-2030 (born FM Esfandiary).
6 Julian Huxley's essay was published in *New Bottles for New Wine* (London: Chatto & Windus, 1957).
7 FM Esfandiary, *Up-Wingers* (New York: John Day Company, 1973).
8 From www.extropy.com/faq.htm (accessed 6 Jan 2006).
9 From author interview, March 2005.

Part 2: Implications, questions and concerns

6. Brain gain

Steven Rose

'Better Brains', shouted the front cover of a special edition of *Scientific American* in 2003.[1] The titles of the articles inside formed a dream prospectus for the future: 'Ultimate self-improvement'; 'New hope for brain repair'; 'The quest for a smart pill'; 'Mind-reading machines'; 'Brain stimulators'; 'Genes of the psyche'; and 'Taming stress'. These, it seems, are the promises offered by the new brain sciences, bidding strongly to overtake genetics as the Next Big Scientific Thing.

The phrases trip lightly off the tongue. There is to be a 'posthuman future' in which 'tomorrow's people' will be what one author describes as 'neurochemical selves'. But just what is being sold here? How might these promissory notes be cashed out? Is a golden 'neurocentric age' of human happiness 'beyond therapy' about to dawn? Will implanted microchips turn our offspring into cyborgs? And if these slogans do become practical technologies, what then? What becomes of our self-conception as humans with agency, with the freedom to shape our own lives? And what new powers might accrue to the state, to the military, to the pharmaceutical industry, to intervene yet further in our lives?

Of course, one is entitled to be a little sceptical. Biological visionaries promising new utopias have been around for many years. Since the 1980s, geneticists have been promising the elimination of disease and the engineering, if not of happiness, then at least of improved health. Some have called this first decade of the new millennium the 'decade of the mind'. Certainly the neurosciences have become growth industries: 30,000 of us meet each year in the US, a more modest 8000 biennially in Europe. With the World Health Organization (WHO) suggesting that psychiatric distress, notably depression, has become a worldwide epidemic, and the incidence of dementias such as Alzheimer's disease increasing inexorably, the pressures to find treatments, probably drug-based, have mounted.

Meanwhile, in the panicky environment of the so-called 'war on terror' there is increasing military interest in the development of techniques that can survey and possibly control and manipulate the mental processes of potential enemies. And as politicians lament the rise of anti-social behaviour and the loss of 'respect' there is a groundswell of interest in the question of whether these social disorders may have biological causes. Might there be neurologically predisposing factors that brain imaging could reveal and pharmacological intervention prevent?

When I entered into it four decades ago, my science was, at its most ambitious, an enquiry into the relationships between brain and mind. More modestly, it attempted to explore the molecular and cellular processes involved in nervous function. Somehow, and quite suddenly, it has moved to the front line of moral and political concern. This is not, by and large, because of new theoretical insights, but instead from technical developments in many areas. Brain researchers are magpies, seizing on methods ranging from genetic manipulation to informatics and imaging in their efforts to comprehend the most complex structure in the known universe. The problems are formidable. Packed into the 1.5 kilos of the human brain are a hundred billion neurons (nerve cells) with possibly a hundred trillion connections (synapses) between them. Most are generated within the nine months from conception to birth, the product during this period and more dramatically in the decades that follow, of the ordered interplay of genes and environment. 'Understanding' the brain requires study over many orders of magnitude from molecules to cells to systems, and ultimately behaviour. It means asking why, and how, brains have evolved in the interests of the organisms, both human and non-human, that possess them. To many of these questions we still have no answers; we are data rich and theory poor. But in the headlong expansion of the last decades, neuroscience has become neurotechnology. What are the promises, threats and implications of this transition? What can we expect over the next two decades?

Curing brains and minds

To begin with the up-side, which most neuroscientists would choose to emphasise, there is now the prospect of treating or even curing brain disorders. The direct threats to the brain are from damage and degeneration. Unlike other body tissues, the central nervous system does not regenerate – hence the disastrous consequence of spinal cord injuries or stroke. For many years now, the neuroscience community has felt itself to be on the brink of solving the problems of regeneration, only to be defeated. The assumption that stem cells derived from human embryos will succeed where earlier attempts have failed is the source of much of the current hype, and has contributed to the UK government's bulldozing aside of the ethical objections felt by many. However, even putting aside these concerns, more than 20 years of experimenting with the use of such embryonic tissue in animal models has so far done little more than reveal the huge problems that need to be solved before human use could reasonably be considered. Even then it may be that adult stem cells, preferably harvested without such ethical dilemmas from the individual who is to be treated, will prove most effective in treating spinal injury and possibly Parkinson's disease. In the light of recent work from the National Institute for Medical Research there are grounds for cautious optimism here.

Alzheimer's, an increasingly familiar disease in an ageing population, is a different story, though also a modestly optimistic one. Despite frequent claims, stem cells are almost certainly irrelevant to its treatment. Genetic and biochemical research has identified the molecular cascade that leads to the neuronal death and cognitive decline that are characteristic of the disease. None of the current generation of drugs are very effective in stabilising – still less reversing – this decline. Novel compounds, based on increased biochemical understanding of the disease, should be available within three to five years, though their use as potential 'smart drugs' poses new dilemmas. Such drugs are likely only to delay an otherwise inexorable decline. The ideal would be to prevent the degeneration rather than ameliorate its symptoms, but that lies further into the future.

When one moves from brains to minds – to psychiatric distress – the picture is a great deal bleaker. The WHO estimates that depression will be the major epidemic disease of our century, affecting up to 20 per cent of the world's population. Women are three times more likely to be diagnosed with depression than men. Mainstream psychiatry tends to ignore this suggestive epidemiology and to 'molecularise' the condition. Successive generations of drugs have culminated in the widespread use of SSRIs (selective serotonin reuptake inhibitors), which boost the function of a particular neurotransmitter, serotonin. Yet despite the publicity that has surrounded them, Prozac®, Seroxat® and their relatives are not much more effective than the drugs they have replaced, and can even be dangerous.

The pharmaceutical industry now pins its hopes on developments in pharmacogenetics, which are based on the idea that people differ in their response to drugs because of small genetic differences. If these differences can be identified, drugs could be individually tailored to match a person's unique genes. However, the rush of optimism about pharmacogenetics that followed the 'reading' of the human genome has been tempered by caution. First, genetic information may not be sufficient in the absence of much greater knowledge of the interactions of genes with one another and with the environment during development. And second, the industry depends on having a mass market for its products; if this market is fragmented by genetic or developmental differences resulting in the need for many different drug types, the economics become trickier.

The mysterious condition known as schizophrenia presents even sharper problems. Although estimated to affect between 0.5 per cent and 1 per cent of the population worldwide, its aetiology is obscure and the claims for genetic causal factors disputed. It affects men and women more or less equally, but working class people are considerably more likely to be diagnosed as schizophrenic than middle class people, and, in the UK, people of Caribbean origin more than those whose biogeographical ancestry lies in Britain. Current drugs are only modestly effective, but a recent Foresight survey[2] of the pharmaceutical industry concluded that there were few prospects for new drug treatments in the immediate future.

Medicalising social problems

One of the most conspicuous features of current social thinking is the tendency to transfer complex social problems to the level of the individual. The person, and not their social context, becomes the focus of treatment. Women are much more likely to be diagnosed as depressed as men, yet this is regarded as a given, not the starting point for questions about gender disparities in power in society. Similarly, the psychiatric disorders listed in the US 'bible', the *Diagnostic and Statistical Manual*, include categories such as 'conduct disorder', 'oppositional defiance disorder' and 'anti-social behaviour disorder'. The paradigm case is 'attention deficit hyperactivity disorder' (ADHD), which supposedly characterises children, mainly boys aged between eight and 13 who are unruly, inattentive and disruptive. Estimates for the prevalence of this condition vary widely: up to 10 per cent in the US for instance. An almost unrecognised disorder in the UK in the 1980s, it is now supposed to affect 1–3 per cent of children here too. The most commonly prescribed treatment is the amphetamine-related drug methylphenidate (Ritalin), which interacts with receptors for the neurotransmitter dopamine in the brain. Drug prescriptions for ADHD in the UK have increased from some 2000 a year in the early 1990s to some 150,000 a year today. Methylphenidate is supposed to make children more tranquil in class and to improve learning. In the US, a diagnosis of ADHD in a child is said to be predictive of delinquent behaviour in adolescence and adulthood. Methylphenidate is thought to reduce this risk.

Such developments highlight the deep ambivalence in modern society towards the use of drugs affecting brain and behaviour, whether legal or illegal, prescription or over the counter. The growing belief in a 'pill for every ill' ignores the ways that a child's discontent at school might be caused by a poor home environment, inadequate teachers, rigid syllabuses or even endemic racism. We seem to be heading towards a pharmacologically defined future, what the neurophysiologist José Delgado called a 'psychocivilised society' and the psychologist BF Skinner offered as 'beyond freedom and dignity'.

On the borderline between curing and enhancing

These developments have also proved a happy hunting ground for the burgeoning profession of bioethics. Is there a moral distinction between treating a deficit and enhancing 'normalcy'? In many ways the question is spurious, hinging on what is meant by 'normalcy' (which has a tricky double meaning, at once statistical and normative). Rectifying short sight with spectacles is treating a deficit, but using a telescope or microscope is an enhancement. It is true that when Galileo developed the telescope there were those among his compatriots who refused to look through it, but few today would share this ethical discomfort. Yet in the context of substances that interact directly with our bodily biochemistry, we feel a considerable unease, reflected in custom and law. It is alright to change our body chemistry by training, but to achieve a similar effect with steroids is illegal for athletes. It is alright to buy educational privilege for ones' children by paying for private tuition, but dubious to enhance their skills by feeding them drugs.

Methylphenidate is a case in point. It is regarded as proper to compel children diagnosed with ADHD to take the drug, but if they trade it in the school playground to a 'normal' youngster wishing to enhance his learning skills, they are condemned. One of the widespread concerns about the development of drugs for the treatment of Alzheimer's is that, because they are aimed at preventing cognitive decline, they are a form of 'smart drug', which will find their way into schools and colleges. Is it cheating to pass a competitive examination under the influence of such a drug? Polls conducted among youngsters make it clear that they do regard it as cheating, in the same way that the use of steroids by athletes is considered to be cheating. However, the military at least has no qualms about such enhancement. US pilots in the recent Iraq war were said routinely to be using the attention-enhancing and sleep-reducing drug modafinil (Provigil) on bombing missions.

Reading our minds, controlling our thoughts

We now live in a surveillance society. On the streets, CCTV cameras monitor our movements. The wonders of information technology ensure that intimate details of our habits and vices are analysed through our use of credit cards, and this information in turn is used to shape our needs and desires through appropriately targeted advertising. But recent developments in the neurosciences are offering more direct access to our most private thoughts, through the new windows into the brain provided by neuroimaging. The extraordinary views of regions of the brain 'lighting up' (albeit in enhanced false colour computer images) when a person is thinking of their lover, imagining travelling from home to the shops or solving a mathematical problem are entrancing. Such imaging techniques (first PET – positron emission tomography; then fMRI – functional magnetic resonance imaging; and now also MEG – magneto-encephalography) began as research tools. Their clinical potential is clear: to be able to identify precisely regions of brain damage, the sites of tumours or the diagnostic signs of incipient dementia.

But what if they could do more? Advertisers and marketers for major companies, ranging from Coca-Cola to BMW, are starting to image the brains of potential customers responding to new designs or brands. New fields of 'neuromarketing' and 'neuroeconomics' are emerging. These trends may be relatively innocuous, but the increasing state and military interest in these techniques is less so. Could brain imaging predict future behaviour? There are claims for instance that such imaging could reveal potential 'psychopathy', that the brains of men convicted of particularly brutal murders show significantly abnormal patterns (although this does not include politicians who send their troops into violent war). In the current legislative climate, where there have been attempts to introduce pre-emptive detention for 'psychopaths' who have not yet been convicted of any crime, such claims need to be addressed critically. They are and will be resisted by the judiciary, but recent developments suggest that this may be a frail defence against an increasingly authoritarian state.

Of course, there are military as well as civil possibilities to consider. In the US (and presumably in the UK as well) such interest goes back at least half a century. A little history is important to put the current situation in perspective. Impressed by claims that the Soviet Union was developing powerful methods of psychological warfare, the CIA and the Defense Advanced Research Projects Agency (DARPA) began their own programmes, recruiting psychologists and neuroscientists to the work. Early experiments included the clandestine feeding of LSD to their own operatives and attempts at 'brain-washing'. These were the forerunners of the hoods and white noise used by the British in Northern Ireland until judged illegal, and of course more recently in Abu Ghraib and Guantanamo.

By the 1960s, DARPA, along with the US Navy, was funding almost all US research into so-called 'artificial intelligence', in order to develop methods and technologies for the 'automated battlefield' and the 'intelligent soldier'. Contracts were let and patents were taken out on techniques aimed at recording signals from the brains of enemy personnel at a distance, in an attempt to 'read their minds'. Primitive at first, these efforts have burgeoned in the aftermath of the so-called 'war on terror'. One company claims to have developed a technique called 'brain fingerprinting', which, according to its website, can 'determine the truth regarding a crime, terrorist activities or terrorist training by detecting information stored in the brain'. We may be sceptical about the validity of such claims but they point clearly to the direction in which such research is currently heading.

The next step beyond reading thoughts is to attempt to control them directly. Once again, there is a long history of attempts by DARPA to develop techniques for focusing microwave beams to disorient or confuse opponents. Whether microwave technology is capable of achieving this goal is uncertain. More promising, however, is a much newer technique – transcranial magnetic stimulation (TMS). This focuses an intense magnetic field on specific brain regions, and has been shown specifically to affect thoughts, perceptions and behaviours that are dependent on those regions. Currently usable only when a subject's head is placed inside the relevant machine, TMS at a distance is now under active investigation. So is chip technology, which might provide implanted prostheses to overcome sensory deficits or control behaviour.

So what should we do about it?

This somewhat telegraphic survey suggests that among the likely benefits to emerge from neurotechnologies there will also be attempts to develop physical techniques for altering mental processes. These include techniques for direct surveillance of citizen's thoughts, which could be used for pre-emptive incarceration or medical treatment. The prospect of these developments raises sharp questions. Neither the science nor the technology (although it is increasingly difficult to make a clear distinction between them) can occur without major public or private expenditure. Their goals are set at least as much by the market and the military as by the disinterested pursuit of knowledge.

Yet they are still only goals, and they are therefore at the point where 'upstream' debate and regulation, as discussed in the Demos pamphlet *See-through Science*,[3] may be effective. They are also at the point – beyond current scientific reality, but not in the realm of science fiction – where the various experiments in public dialogue that have been tried in many European countries in the past decades – deliberative democracy, citizen's juries and the like – could have a part to play. Because the science and technology are international, so too must be these attempts to develop effective, rather than sham, forms of participation. Neuroscientists, in this context, have a responsibility to make their subject and its potentials as transparent as possible. As I write, a unique attempt in this direction is being made in the pan-European Meeting of Minds project, coordinated by the King Baudouin Foundation in Brussels, due to present its findings to the European Parliament in early 2006. Whether the voices of these citizens of nine European countries will be listened to in the cacophony of slogans about 'better brains' remains to be seen.

Steven Rose is Professor of Biology and Director of the Brain and Behaviour Group at the Open University. His most recent book is The 21st Century Brain *(London: Jonathan Cape, 2005).*

Notes

1 'Better Brains?', *Scientific American*, Sep 2003.
2 *Foresight Drugs Futures 2025?* Perspective of the pharmaceutical industry (London: Office of Science and Technology, July 2005).
3 J Wilsdon and R Willis, *See-through Science: Why public engagement needs to move upstream* (London: Demos, 2004).

7. The cognition-enhanced classroom

Danielle Turner and Barbara Sahakian

'Smart drugs' are used by all sectors of society to improve the functioning of the human mind. But there is now growing evidence, particularly from the United States, that pharmaceuticals are being both prescribed and illegally consumed by university students to maintain supernormal levels of concentration in the run-up to exams, with the suggestion that this trend will eventually encompass younger children. How should society react to this increasing desire by people to use smart drugs? What effects could their widespread use have on our educational systems? Could children in the future face blood or urine tests when sitting their A-level or GCSE exams?

Recent developments in drugs to improve memory and cognition certainly raise the prospect of drug-testing regimes in schools similar to those imposed on athletes. It is essential that educators in particular think hard about the implications of such developments. Are the smart drugs of the future more likely to be viewed as giving an unfair advantage to pupils, or will they be embraced by parents and teachers as a reasonable addition to the armament of self-improvement techniques designed to give children the best possible start in life?

Until recently, psychotropic medications had significant risks that made them attractive only when the benefits to the patient were considered to outweigh the side effects. However, it is now becoming possible to enhance cognition pharmacologically with minimal side effects in healthy volunteers. For example, as part of a research programme to identify cognitive enhancers for patient use, we showed in our laboratory in Cambridge that a single dose of modafinil (Provigil, a drug licensed for the treatment of narcolepsy) induced reliable improvements in short-term memory and planning abilities in healthy adult male volunteers.[1] Improvements in performance have also been shown in healthy young male students after a single dose of methylphenidate (Ritalin).[2] Some research has indicated similar cognitive-enhancing potential with a group of memory-modulating drugs called ampakines.[3]

Such drugs are typically developed to treat a medical condition, but are proving to be safe enough for widespread use following healthy volunteer studies. The list of agents, including nutraceuticals and herbal enhancers, is also growing.[4] More work is needed to determine if these drugs will maintain their beneficial effects when taken over a long period of time. Nevertheless, in the absence of contrary advice, increasingly they will be used for indications other than those they are licensed for.

The use and abuse of prescription drugs

Most of the evidence for off-label use of smart drugs by students and young adults currently comes from the United States. Researchers at the University of Michigan showed recently that just over 8 per cent of university undergraduates report having illegally used prescription stimulants.[5] The most common motives given by students for the use of such stimulants are to help with concentration and increase alertness, followed by a desire to get high. These findings are backed by reports from the National Institute on Drug Abuse in the United States that, in 2004, 2.5 per cent of eighth graders (approximately 13–14-year-old children) abused methylphenidate, as did 3.4 per cent of tenth graders and 5.1 per cent of twelfth graders.[6] A separate but equally burgeoning phenomenon is of students obtaining prescriptions for stimulants through diagnosis of conditions such as attention deficit hyperactivity disorder (ADHD).

In the United States it is estimated that almost 700,000 doses of methylphenidate were stolen between January 1996 and December 1997, with 15 per cent of students using illegal stimulants thought to be obtaining the drugs through theft.[7] This is likely to stem from the difficulties that healthy individuals encounter in their attempts to obtain prescription drugs. Currently in the UK (and the US) there is no regulatory framework in place to enable the licensing of drugs for use in healthy individuals. Drugs are either licensed for medicinal use in patients via the Medicines and Healthcare products Regulatory Agency or controlled under the Misuse of Drugs Act. Smart drugs are most likely to be obtained illegally via the internet or with a private prescription from a sympathetic prescriber. It is unlikely that there will be a regulatory change regarding drugs for people who have not been diagnosed with a psychiatric illness. Fear of litigation means that pharmaceutical companies developing smart drugs for use in clinical groups are not keen to seek a licence for these drugs to be used by healthy individuals. Nevertheless, some prescription drugs can be more readily obtained than others because they are licensed for more broadly defined illnesses. For example, the licence for modafinil was recently extended to include the condition of excessive daytime sleepiness, potentially opening an avenue for many more people to obtain this drug under broader diagnostic criteria.

What counts as enhancement?

There are many difficulties in defining what should be considered 'normal'. The subtleties of modern medicine, combined with the expectations of a well-educated public, mean that the distinction between treatment and enhancement is often blurred. In practice many conditions (including ADHD) present as spectrum disorders with a grey area in which diagnosis is largely subjective. It is impossible to determine categorically whether a child or student is functioning within the 'normal' range, or is suffering from a psychiatric condition requiring treatment. For example, despite attempts at standardising diagnostic criteria, cross-cultural studies of symptoms of ADHD show significant differences in the diagnosis of childhood ADHD across different countries, in that of children from different cultures within the same country, and even of children from within the same culture by different diagnosticians.[8]

Furthermore, there are anecdotal reports of children younger than three years old (the current licensing limit) being prescribed stimulant medication for ADHD. Difficulty in diagnosis at such young ages increases the likelihood that children are receiving unnecessary drug exposure. Differing social and philosophical opinions make it difficult to determine what should be considered a sufficient impairment to warrant pharmacological intervention. However, scientific advances in objective biomedical markers, at least, are likely to improve diagnostic accuracy in the future to ensure that those children most in need of help will receive it.[9]

In addition to questions relating to the definition of 'normal', there are additional concerns about the safety of the use of smart drugs. This is particularly true if a pharmacological agent is to be used to enhance, rather than to treat. Is it ethical to make available drugs that potentially could cause harm to healthy individuals? It is always difficult to be certain about the potential for subtle, rare or long-term side effects, particularly in relatively new pharmaceuticals. Children, especially, are at risk from drugs that could adversely affect brain development. For example, researchers at Harvard Medical School showed recently that administration of methylphenidate to adolescent rats results in long-lasting behavioural changes and molecular alterations in the function of the brain's reward systems.[10]

Weighing up benefits and risks

Despite the difficulties inherent in monitoring the effects of drug usage over several years, a full exploration of the long-term implications of new treatments is vital, especially those that might routinely be used by the healthy population. Pharmaceutical companies and drug regulators already invest considerable resources in ensuring the safety of drugs, although most of the safety studies are undertaken in adult groups and not child populations. Nevertheless, many believe that there is considerable underreporting of adverse drug reactions by healthcare professionals in the UK and that harmful drugs could be identified sooner.[11] Strategies are being put in place to increase early identification of harmful drugs, including encouraging patients – as well as healthcare professionals – to report adverse drug reactions, and providing a publicly available global clinical trials register aimed at ensuring that the results of all pharmaceutical research trials (including 'in-house' studies) are disclosed.

No drugs are side-effect-free, which means there is a need for risk–benefit analyses that specifically consider the use of drugs for enhancement rather than treatment. This is especially true in paediatric care. With the advent of pharmacogenomics – the discipline behind our increasing understanding of how genes influence the body's response to drugs – it is likely that the risk of side effects can be considerably reduced. It is also important to remember that the effects of smart drugs are not homogeneous, nor entirely predictable. For example, in healthy young university undergraduates, our laboratory showed that the cognitive-enhancing effects of methylphenidate were limited to when the volunteers were in a novel situation, with no effects being seen when the psychological tasks were familiar to the volunteers.[12] It is also known that improvements in performance may depend on the individual's baseline level of performance. In another study from our laboratory it was found that volunteers with the poorest memory capacity showed the greatest improvement on methylphenidate.[13] Similarly, cognitive-enhancing drugs do not improve all aspects of cognition equally. A single dose of modafinil improves short-term memory and planning abilities, but has no effect on the ability to sustain attention in healthy individuals. Methylphenidate, in contrast, primarily affects attention. People might thus have to take several different cognitive enhancers to target all the functions they want to improve, with a risk of drug interactions and increased side effects.

Drugs in the classroom

If educators are to make decisions about the use of smart drugs by students and school children it is important to examine the reasons behind their use. If students feel compelled to take cognitive enhancers in order to improve their abilities to concentrate, are they simply succumbing to the intensifying demands of a 24/7 society? Are unrealistic feats of memory and attention being expected of today's students? Are parents demanding drugs for their children in order to help them succeed against increasing numbers of medicated contemporaries?[14] Or are the main pressures from schools and teachers desiring better-behaved classrooms? Should education systems be restructured towards guiding students to lead fulfilling, responsible lives as adults, instead of being driven primarily by exam results? And if this were the case, would we see the same phenomenon of children and students resorting to pharmacological solutions to their difficulties?

There are also questions about the more intangible effects smart drugs could have on children and students. Is it possible that these drugs could be used to reduce social inequality and injustice in society? Or it is more likely that their use will fuel further disparity based on a lack of affordability? Could cognitive enhancers have unexpected social ramifications, as people are deprived of a sense of satisfaction at their own achievements? How likely is it that human diversity could be limited through the widespread use of these drugs?

As our scientific understanding advances, there is a need for educators, the government, academics and the public to start an open debate about these issues. One recent proposal is for the creation of professional 'neuroeducators', who could guide the introduction of neurocognitive advances into education in a sensible and ethical manner.[15] Already a number of UK universities, including Cambridge, are offering courses that consider neuroscience in education. However, a new cadre of neuroeducators should not be expected to provide answers to all of the ethical dilemmas posed by smart drugs and other advances. Children have the right to an open future, and a delicate balance must be struck between an individual's right to use psychoactive substances, their responsibilities to society, and indeed society's responsibility to the individual.

Danielle Turner is a research associate and Barbara Sahakian is Professor of Clinical Neuropsychology in the Department of Psychiatry at Cambridge University. Danielle Turner is also a fellow of the Centre for Cognitive Liberty and Ethics (USA).

Notes

1 DC Turner et al, 'Cognitive enhancing effects of modafinil in healthy volunteers', *Psychopharmacology (Berl)* 165 (2003).
2 R Elliott et al, 'Effects of methylphenidate on spatial working memory and planning in healthy young adults', *Psychopharmacology (Berl)* 131 (1997).
3 G Lynch, 'Memory enhancement: the search for mechanism-based drugs', *Nat Neurosci* 5 Suppl (2002).
4 R Jones, K Morris and D Nutt, *Cognition Enhancers* (2005). Available at: www.foresight.gov.uk/Brain_Science_Addiction_and_Drugs/Reports_and_Publications/ScienceReviews/Cognition%20Enhancers.pdf (accessed 4 Jan 2006).
5 CJ Teter et al, 'Prevalence and motives for illicit use of prescription stimulants in an undergraduate student sample', *J Am Coll Health* 53 (2005).
6 NIDA InfoFacts. Methylphenidate (Ritalin). *National Institute on Drug Abuse* (2005). Available at: www.drugabuse.gov/pdf/infofacts/Ritalin05.pdf (accessed 4 Jan 2006).
7 E Kapner, 'Recreational use of Ritalin on college campuses', *InfoFactsResources – The Higher Education Center for Alcohol and Other Drug Prevention* (2003). Available at: www.edc.org/hec/pubs/factsheets/ritalin.pdf (accessed 4 Jan 2006).
8 FX Castellanos and R Tannock, 'Neuroscience of attention-deficit/hyperactivity disorder: the search for endophenotypes', *Nat Rev Neurosci* 3 (2002).
9 Academy of Medical Sciences, *Safer Medicines: A report from the Academy's FORUM with industry* (November 2005). Available at: www.acmedsci.ac.uk/images/page/1132655880.pdf (accessed 4 Jan 2006).
10 WA Carlezon Jnr and C Konradi, 'Understanding the neurobiological consequences of early exposure to psychotropic drugs: linking behavior with molecules', *Neuropharmacology* 47 (2004).
11 National Audit Office, *Safety, Quality and Efficacy: Regulating medicines in the UK*. Report by the Comptroller and Auditor General HC 255 (London: Stationery Office, 2003).
12 Elliott et al, 'Effects of methylphenidate'.
13 MA Mehta et al, 'Methylphenidate enhances working memory by modulating discrete frontal and parietal lobe regions in the human brain', *J Neurosci* 20, no RC65 (2000).
14 I Singh, 'Will the "real boy" please behave: dosing dilemmas for parents of boys with ADHD', *Am J Bioeth* 5 (2005).
15 K Sheridan, E Zinchenko and H Gardner, 'Neuroethics in education' in J Illes (eds), *Neuroethics: Defining the issues in research* (New York: Oxford University Press, forthcoming).

8. Better by design?

Sarah Franklin

3 May 2003. The glossy cover of the *Guardian Weekend* magazine features a provocative image of a sonogram of a fetus reading a volume of Proust to accompany an article by Bill McKibben warning of the dangers of the designer baby era.[1] In provocative visual language, the image conveys the idea of a new era of genetic manipulation and made-to-order, 'superior' offspring.

Like the figure of the clone, the designer baby has become an iconic signifier of the dilemmas and risks posed by new genetic technologies. The cover is headlined: 'Condemned to be Superhuman: the terrifying truth facing tomorrow's babies', leaving little room for doubt that the designer baby is a threat to humanity. This popular view is not only a ubiquitous media shorthand for genetic science going too far but is equally prominent in recent books by leading public intellectuals, such as Francis Fukuyama[2] and Jürgen Habermas.[3]

Both authors identify preimplantation genetic diagnosis (PGD) with 'designer babies'. As Fukuyama writes: 'In the future it should be routinely possible for parents to have their embryos automatically screened for a wide variety of disorders, and those with the "right" genes implanted in the mother's womb.' Based on such predictions, he warns of an inevitable collapse of the boundaries between genetic screening, diagnosis and enhancement. Fusing the imagery of information technology ('a few clicks of the mouse') with that of 'routine' genetic profiling, he imagines a scenario of embryos being 'automatically analysed' and 'enhanced'. Importantly, it is PGD that provides the crucial interface between reproduction and genetics for Fukuyama, by offering parents 'the first step toward . . . greater control over the genetic make-up of their children'.[4]

Similarly, Jürgen Habermas warns that the future of human nature is imperilled by the kinds of genetic choices PGD enables. In *The Future of Human Nature*, he claims that: 'genetic manipulation could change the self-understanding of the species in so fundamental a way that the attack on modern conceptions of law and morality might at the same time affect the inalienable normative foundations of societal integration'.[5]

This view of genetic manipulation as *a force unto itself*, hostile to social order and integration, is repeated in many of Fukuyama's and Habermas's dire predictions about an 'automated' genetic future for the human species. As Habermas claims:

> The deepest fear that people express about technology *is . . . that, in the end, biotechnology will cause us in some ways to lose our humanity – that is, some essential quality that has always underpinned our sense of who we are and where we are going. . . . Worse yet, we might make this change without recognising that we had lost something of great value.*[6]

Here, again, 'biotechnology' is attributed a sinister agency: it is in our hands but we *might not even recognise* its potential to change our very nature. Writing in the *Guardian Weekend* in a similar vein, Bill McKibben urges his readers to pause to consider 'the terrifying truth' ahead of us, and to decide to act – to draw the line against designer enhancements. Describing 'where we are', he claims that:

> *The genetic modification of humans is not only possible, it's coming fast: a mixture of technological progress and shifting mood means it could easily happen within the next few years. But we haven't done it yet. For the moment we remain, if barely,*

a fully human species. And so we have time to consider, to decide, to act.[7]

Like Fukuyama and Habermas, McKibben depicts a process that is *out of control*, and in need of restraint. He attributes the 'terrifying' forward march of genetic modification to what he calls 'a combination of technological progress and shifting mood' and argues it is time to call a halt. We are becoming, in the words of his article's title, 'too clever, too fast, too happy'. What is sinister, then, is not only the *nature* of the change at hand, but the *process* by which it is occurring, without our consent, because we do not even notice. While scientists are busy producing a 'super-race', the general public remains unaware of the magnitude of the powers they have unwittingly allowed to threaten our species' identity.

This popular but pessimistic view of scientists – as powerful, untrustworthy and socially marginal – positions them as both ahead of, and outside or beyond, social norms, while they increasingly control forces almost too terrifying to contemplate. It is a view closely associated with the image of Dr Frankenstein secretly creating a monster in his isolated laboratory. This stock characterisation emphasises a dramatic separation between scientists and society, and a conflict of interests between them. It exaggerates the remoteness of scientific knowledge, and its disconnection from 'ordinary life'. While 'we' are getting on with business as usual, 'they' are designing made-to-order babies. Above all, it is a view that emphasises secrecy.

McKibben's scandalised and denunciatory rhetoric is closely aligned to that of Fukuyama and Habermas, then, not only through their joint call to halt the sinister, 'unnoticed', forward march of cloning and genetic manipulation, but in their shared emphasis on the undesirability of genetic design, and the extent to which it is already routinised, automated, and out of control. The message from all three of these prominent authors is that the 'designer' revolution in genetics is moving ahead too quickly, without our consent, and that it is not something we really want because it will destroy who we really are.

Desires and designs

Yet other commentators on the 'genetic design' debate take a different view.

One prominent voice is that of bioethicist Gregory Stock,[8] who argues that far from being 'some cadre of demonic researchers hidden away in a lab in Argentina trying to pick up where Hitler left off', the scientists and clinicians who are laying the foundations for 'the reshaping of genetics and biology' are pursuing 'mainstream research that virtually everyone supports', including treatments for infertility, which 'is a source of deep pain for millions of couples'.[9] In the conclusion to *Redesigning Humans*, Stock claims that rejection of new reproductive and genetic technologies is not only a misguided mission but a redundant one:

> *There is no way we can permanently forego these enhancement technologies if they prove robust and useful. Those who shun healthier constitutions and extended lifespans might hope to remain the way they are, linked to a human past they cherish. But future generations will not want to remain 'natural' if that means living at the whim of advanced creatures to whom they would be little more than intriguing relics from an abandoned human past.*[10]

Stock's advocacy of 'greater germinal freedom' stands in sharp contrast to the views of Habermas, McKibben and Fukuyama in that he emphasises the broad base of popular support for scientific intervention into reproduction, and the impossibility of separating biomedical innovation from treatment for conditions such as infertility.

But why are these arguments so focused on the idea of 'design'? Why is 'designer' equated with genetic 'modification'? Technically, the only difference between PGD and *in vitro* fertilisation (IVF) is that embryo selection is based on genetic information and morphology, instead of just morphology alone. And to be able to diagnose the presence or not of a known, single and specific mutation is not the same as modifying it. In addition to the famously imprecise concept of 'human nature', a major area of confusion in the commentaries discussed above is the extent to which desire, demand and design are constantly conflated in the depiction of genetic modification as 'out of control'.

McKibben, Habermas and Fukuyama view genetic 'modification' as a sinister force that is racing ahead 'unnoticed', driven by scientific desires, whereas Stock, and other prominent scientists including James Watson, argue that nothing could be more obvious than that parents would want to 'apply the benefits of discovery' to provide 'our very best for our children'. But what, exactly, is the agent of change, and what is the change itself? Does it lie in the inexorable forward march of technology, as McKibben suggests? Or is it to be found in biologically determined parental desires? Is human nature the driving force behind technological innovation? Or is human nature itself at risk of being eliminated by the 'automatic' and unstoppable progress of science and technology? Habermas warns that 'the deepest fear that people express [is that] biotechnology will cause us . . . to lose our humanity'.[11] But such warnings are confusing: is something taking our humanity away from us? Or is this something we are doing to ourselves?

Whether the terms 'design' or 'designer' are deliberately obfuscating, simply a shorthand for PGD, part of the usual media hype (or some combination of all of the above), the designer baby has become a highly contested nexus of conflicting opinion, much of which is confused, contradictory and ambivalent. Beginning with the more 'high-brow' commentaries of philosophers, journalists, scientists and bioethicists we see a range of divergent certainties, from the belief that the designer baby will be our undoing, to the conviction it is not in the least threatening, but rather as natural as the urge to parent itself. Such accounts reflect three significant tendencies in public debate about designer babies – be it in Britain, Germany or the United States. These tendencies are:

1 to depict PGD as a mixture of desire and design

2. to position PGD as a threshold technology or an interface to an improved or degraded future
3. to express ambivalence, confusion and equivocation about the 'designer baby' technique in terms of its future consequences.

Thus, although use of the terms 'designer' and 'design' to refer to PGD is technically inaccurate (as no design or modification is involved), and may be described as misleading and harmful (as many PGD patients do), they are still important from a cultural or sociological point of view as 'placeholders' for issues that may be difficult to explain, or even articulate. It is precisely because the term 'designer' is both so vague and so ubiquitous that it is worthy of further investigation.

Governing the future of PGD

To understand how PGD came to occupy such a pivotal position in contemporary debate about 'designer' reproductive futures, it is useful to return to its beginnings. PGD was developed scientifically in the UK, where it has played an important and distinctive role in public debate, and in the process of devising legislation. One of the striking features of this history is the role of PGD in focusing and clarifying public attitudes towards reproductive biomedicine. Consequently, one of the most important parts of the PGD story in Britain is the decisive role of this technique as early as 1985 in the elaborate process of devising legislation to govern 'human fertilisation and embryology' – a process that took more than a decade to complete, and which is ongoing. PGD's significance to this history derives in part from its transformation from being a scientific possibility into a clinical reality during exactly the time period of legislative 'gestation' of the Human Fertilisation and Embryology Act, namely 1984–1991. Another striking feature of the PGD debate in Britain is the unusually prominent role of scientists and clinicians, and in particular embryologists, in both parliamentary and public debate.

The birth of Louise Brown in 1978 was seen to create not only a new kind of reproductive choice, but a legal vacuum surrounding its use, as well as an immediate practical imperative to produce regulation. A resolution to this legal and regulatory challenge was both lengthy in its evolution and comprehensive in its scope: Britain's renowned Human Fertilisation and Embryology Act of 1990 emerged 12 years after the world's first test-tube baby was born in Oldham, Lancashire. Its enactment concluded an unprecedented process of public consultation and parliamentary negotiation. The Act remains the most extensive, substantial and detailed legal framework ever created to regulate and govern what had previously been the legally uncharted territory of 'human fertilisation and embryology'.

Since 1990, the Act has been copied by countries all over the world and is widely seen as a unique, exemplary and distinctively 'British' achievement that continues to set the global standard for governance of the post-IVF reproductive 'revolution'.

However, IVF was not the only technique to play a leading role in the battle to establish governance over 'human fertilisation and embryology'. PGD played an equally influential role during the passage of the Human Fertilisation and Embryology Act, and its importance to the shape of contemporary regulation and debate of reproductive biomedicine in Britain has steadily increased, even to the point, in 2005, of precipitating a wholesale review of the Act, and the creation of the Human Fertilisation and Embryology Authority (HFEA), which oversees reprogenetic governance in the UK.

Thus, despite the numbers of actual PGD cycles and patients in the UK remaining in their hundreds, dwarfed by the vast demand for IVF, PGD has, from the mid 1990s onward, become such a pivotal technique, linking IVF to cloning and human embryonic stem cell research, that it is, more than any other issue, the source of questions driving changes to science, governance and policy. For these reasons and others, the 'UK PGD story' provides an essential background to the scientific potency, clinical urgency and political volatility of the 'designer baby' question.

Rather than science 'racing ahead' of society, it is the sociality of science that emerges from this quick sketch of the 'the birth of PGD' in the UK. Rather than being a threat to the future of human nature, or evidence of an attempt to redesign or 'manipulate' human origins, PGD emerges as the centrepoint of a complex political history marked by unpredictable twists and turns, and variously aligned sets of interests. And the establishment of the HFEA demonstrates the capacity for science to be regulated through an extensive and far-reaching process of social negotiation and policy innovation which is ongoing.

Above all, in response to the question 'What is PGD?', it is possible to identify a number of distinct but overlapping analytical dimensions, from the technical to the legislative, and from the clinical to the philosophical. All of these demonstrate how the Warnock Committee's strategy of promoting scientific progress subject to strict regulation has repeatedly set the course for 'the British way forward' in the field of reproductive biomedicine. From this point of view, it is difficult to describe PGD as anti-social, or to recognise it in the stock elements of the 'blue-eyed, blonde-haired designer baby' debate.

Contrary to Habermas's claim that 'people's deepest fears' about new reproductive and genetic medicine are that they will lose their humanity and forfeit the ability to determine their futures, what emerges from a brief scan of PGD and its future is the extent to which PGD in the UK, like stem cells in the US, is associated with public debate and regulation, *not their absence*. Part of the value of examining the history of PGD in specific national contexts is the ability to learn through comparisons, such as those that might be made with any number of countries in which PGD has had a distinctive historical profile, such as Australia, Belgium, Cyprus, Denmark, India, Israel, Russia, South Korea or Sweden. This alone will not settle the ongoing, and often unanswerable, questions posed by new combinations of technological potential, parental desire and the widely shared view of the need for limits to reproductive and genetic intervention. Neither will it be sufficient to point to the wide range of different strategies for regulating techniques such as PGD in the effort to suggest it is in fact society that is leading technology along a consensual path of greater relief of human suffering. The effort to present a 'thicker' account of how people navigate the issues surrounding the use of PGD offers the possibility that some of the terms on which PGD has already been debated might play a larger role in shaping its direction in the future. However, it will not eliminate the gaps between these views and others, which is fortunate, as disagreement may be one of the most important resources ensuring an ongoing debate.

Sarah Franklin is Professor of Social Studies of Biomedicine and Associate Director of BIOS at the London School of Economics. This essay is an extract from her forthcoming book, co-authored with Celia Roberts, Born and Made: An ethnography of preimplantation genetic diagnosis *(Princeton, NJ: Princeton University Press, Nov 2006).*

Notes

1 The article was an extract from McKibben's book *Enough: Genetic Engineering and the End of Human Nature* (London: Bloomsbury, 2003) on the eve of its British publication.
2 Francis Fukuyama is Professor of Political Economy at John's Hopkins University, and a member of the National Bioethics Advisory Council appointed by George Bush and chaired by Leon Kass.
3 Jürgen Habermas is a philosopher who works in the tradition of critical theory; he initially delivered his thoughts on designer babies in response to a lecture by Peter Sloterdijk on 'the human zoo'.
4 F Fukuyama, *Our Posthuman Future: Consequences of the biotechnology revolution* (New York: Farrar Straus Giroux, 2002).
5 J Habermas, *The Future of Human Nature* (Cambridge: Polity Press, 2003).
6 Ibid.
7 B McKibben, 'Condemned to be Superhuman: the terrifying truth facing tomorrow's babies', *Guardian*, 3 May 2003.
8 Gregory Stock is the director of the Program on Medicine, Technology, and Society at the UCLA School of Public Health.
9 G Stock, *Redesigning Humans: Our inevitable genetic future* (New York: Houghton Mifflin, 2002).
10 Ibid.
11 Habermas, *Future of Human Nature.*

9. More life

Jon Turney

'Death is no different whined at than withstood.' This is Philip Larkin, whose work returns again and again to his dread of oblivion. Larkin, my edition of the *Collected Poems* tells me, was the best-loved poet of his generation. I love him too. But does his gloomy protest about the inevitable end of life tell us anything special about our culture? No, death has preoccupied us throughout history. What may be special about our times, though, is the number of people who seem determined to do something about it.

Human beings are the only creatures endowed with both an awareness of death and an imagination. So the stories we tell to console ourselves invariably offer the possibility that death can be evaded or transcended. Logically, this might happen in a number of ways. If there is a soul, ghost or spirit separate from the body, maybe it will survive in an immaterial realm, or find a new host through reincarnation. There might be some trick which negates ageing and allows the body to carry on – a form of physical immortality. There might be resurrection, through supernatural agency. Or, failing any of these, we may simply contribute to posterity and be 'remembered', either genetically or culturally.

To deal with death, a mixed strategy seems best. A good twenty-first century hedge might be to live as a devout Christian who raises children, writes essays and works in a lab researching cures for ageing. Such a person would be maximising the chances of their genes, their ideas and even their body surviving, while believing that God will ensure eternal life when Earthly striving ceases.

I daresay there are people who do all these things. But you cannot help noticing the current popularity of the notion that physical immortality, or at least radical life extension, is the one to bet on. Books on the topic abound. Biotech companies with names like Elixir, Chronogen and Juvenon raise venture capital for ultimate medicines. Newspaper columnists debate whether it could really be true that the first human who will live to be 300, or even 1000, has already been born.

A brief history of immortality

Why now? A historical sketch suggests that this is where modernity has taken us in the West. The ancients, lacking plausible technologies, told stories about immortals but set their sights on the afterlife. That promise carried over into Christianity, in which original sin lost us the infinite life of the Garden of Eden, but gave us a chance of redemption after death. The enlightenment, and the science it brought in its wake, fashioned a new narrative of progress, and of a human paradise (re)created on Earth through collective effort sustained down the generations. Over time, science weakened the faith which underpinned the afterlife, but the rise of evolutionary thought reinforced the notion that improvement was possible. But the new, secular faith in social progress also faded in the twentieth century, leaving only the hope for fulfilment during the individual life. So let it last as long as we can make it.

The final ingredient is that this shift in belief goes along with a new-found sense of scientific possibility. If physical immortality is the only remaining option for denying death's dominion, it also looks more attainable. No more snake oil and monkey glands. We have molecular genetics and, maybe soon, nanotechnology. Our society has already stretched average life expectancy. Surely we will, we must, go further?

I think this sketch is basically sound. Look, for example, at Robert Ettinger's 1965 book *The Prospect of Immortality*, the first proposal for extending life which was inspired by serious, real-world science.[1] OK, the science is a pretty generous extrapolation of some experiments with freezing organisms or parts of organisms – the real breakthroughs are assumed to lie in the future, when the preserved dead will be revived from their 'dormantories'. Still, Ettinger's is a technically informed recipe for resurrection: 'Most of us now breathing have a good chance of physical life after death – a sober, scientific probability of revival and rejuvenation of our frozen bodies,' he declares.

But his promise extends beyond immortality. This scientific, rather than supernatural, resurrection will nevertheless take place in something like Paradise. It will contain, for example, 'intelligent, self-propagating machines', which will 'scoop up earth, or air, or water, and spew forth whatever is desired in any required amounts', as well as repairing themselves and improving their own design. (A couple of decades later, this fantasy machine would become a nanotechnological assembler, but its function is the same.) A materialist Paradise, then, but one which Ettinger suggests is enough to tempt anyone who is unsure about being frozen. 'Before long nearly everyone will see the Golden Age shimmering enchantingly in the distance, and will not dream of relinquishing his ticket.'

This is heady stuff for an author who offered a 'sober, scientific' prospect! But it fits the notion that immortality as a technological project is a substitute religion for a secular, scientific age. The next literary landmark, Alan Harrington's *The Immortalist*,[2] made the idea even more explicit. Harrington's fascinating book, first published in 1969, has a much wider range of cultural and philosophical reference than Ettinger's, and was an influential text for the later transhumanist movement. But the message is basically a simple one. He begins, 'Death is an imposition on the human race, and no longer acceptable.' The answer is to face up to the inadequacy of religion, and take the matter into our own hands using the science we now know can be ours. 'Our new faith must accept as gospel that salvation belongs to medical engineering and nothing else.' An 'immortality program' would not be nearly as expensive as the Manhattan project or the moonshots.

So here we have the beginnings of what has since become a more common view. Turning away from religion and individual salvation through technology becomes a serious alternative. You could almost imagine a state-sponsored effort along the lines Harrington proposed. Not long after he wrote, President Nixon declared a 'war on cancer', inspired by the Apollo programme (it is a war we still seem to be fighting). Why not a war, not just against one dread disease, but against death itself? Enlist, and you have a chance not merely of contributing to posterity, but of witnessing it.

Yet this is not quite the whole story. We are now hearing more often that an effort to achieve physical immortality is a historic necessity. But some have always hoped that it *might* be possible. As Lucian Boia concludes in his engaging survey of ideas about longevity, *Forever Young*,[3] all the ideas about ways to overcome death have been present since early in recorded history. And along with rage against the dying of the light, fables of immortality and trips to paradise, we find the Senecan tradition of Stoic wisdom, and stories – as popular now as then – of the ennui of eternal life, or the horrors of extended decrepitude.

Ending the blight of involuntary death

The fact that such a full spectrum of views has endured for so long suggests that the position one takes now is less logical than temperamental. And a reading of some of the more recent advocates of life extension tends to confirm this. They regard death as an affront, a design flaw and a challenge to be overcome through human ingenuity. And the hint of longer life becoming a real technological project lends an urgency to their rhetoric. They would really hate to be one of what Damien Broderick, the Australian science-fiction author and futurist, describes as *The Last Mortal Generation*.[4]

But it is not just the thought of missing out on a prize now within our grasp which drives the modern immortalists. There is a consciousness of death as the ultimate deadline, of the brevity of life, which precedes this new hope. How can we account for the rise of this particular disposition? I daresay it is one which has always existed, but our culture does seem to have given it greater prominence. Perhaps a useful way to explain is to be personal. After all, death is something everyone has some attitude to, even if it is denial.

Forget immortality, which raises a host of fascinating issues but remains purely speculative. A more psychologically realistic question, I suggest, is to ask: given good health and sound mind, would you like to live another 20 years? This is a span most of us can imagine.

As it happens, I turn 50 in the month when I am writing this. My answer to the 20-year question is a loud yes. I don't exactly have a plan, but there are things I wish to do and see over the next two decades (the details don't matter, it matters that I want to be here to do them).

Come back and ask me the same question in 20 years' time, and my best guess is that I will say 'yes' again. Ask me a third time, at 90, and I am not so sure. But assume another affirmative, and ask me again when I am 110 in 2065. I can well imagine not caring much one way or the other. At the moment, I certainly doubt that I would keep on saying yes until the end of the century.

Do I underestimate my putative 110-year-old self's lust for life? Perhaps, but assume that the way I feel now does indicate how I might feel then. I do not know how common this disposition is, though it seems quite common among the comfortably off British folk that I know. But I believe it contrasts sharply with the advocates of life extension. They are clearly framing their lives in a different narrative. I have watched my children being born and my parents die, but I still don't feel, with Ray Kurzweil, that 'disease and death at any age is a calamity'. I am absolutely gripped by the idea that humans, in some form, will live for many millennia to come, but I don't find the idea of personal finitude an affront. To tell the truth, I don't feel old, but I do feel I have been around quite a long time already. Fifty years is a vanishingly small instant on a geological or cosmic time-scale, but days are where we live (Larkin again), and it adds up to 18,000 of those.

I claim no virtue for this position, merely that it differs sharply from those who think that the current life span is not nearly long enough. If they are more vocal, and their voices more insistent, are they prompted by some other feature of our culture than secularisation and technological prowess? Many in the West now live longer, but seem to feel short of time. We are offered an extraordinary volume of cultural product, as well as a repackaging of 'experiences' both as the moments which make life worthwhile, and as items that can be bought and traded. Somewhere in this culture of 24-hour information, with boxed sets of all the TV shows you never watched the first time round, and a profusion of lists of '100 things you must do before you are 30', we live our everyday lives, cash rich but time poor. And we are told that the path to happiness lies in 'living life to the full', whatever that may mean.

How much is enough?

So if the wish for longer life is part of the human condition, late capitalism works on that wish in its own unstoppable way. The critique is familiar. We clock in and clock out of work, but our time off work is redefined as 'leisure' and colonised by leisure industries. Their need to shift product means they must try and perpetuate our juvenile craving for novelties, or sell us substitutes for non-commodifiable goods like community or conversation.

At the same time, the culture surrounding us is, in its way, astonishingly rich and exciting. We create more now, if only because there are so many humans alive; we recover more of the past; we know more of other cultures. And everything is available to all, all the time. Again, we come up against the human incapacity to sample more than a minute fraction of the world.

So are the advocates of life extension oppressed by lack of time? They do not always say so directly, but you can see it in occasional asides. James Hughes, for example, considering that he is executive director of the World Transhumanist Association, writes a commendably balanced account of the prospects for human enhancement in *Citizen Cyborg*.[5] But he gives a glimpse into what drives him when he suddenly declares that he has always resented sleep, as a time when nothing happens, a waste of a third of one's life.

If any lapse in consciousness is a flaw, then the final dissolution of consciousness is a looming calamity. But if you regard sleep as a rather pleasant part of the day, or even wakefulness as a potentially rewarding interlude between naps, then Hughes' declaration may seem puzzling. No knitting up the ravelled sleeve of care for him.

All the various ways of dealing with mortality, possibly excepting stoicism, can lead to zealotry. But the difference today is that for the first time those pressing for research on life extension may actually be able to implement their programme in a way which alters the facts of the matter. This is one respect in which the meek certainly will not inherit the Earth.

So if the prospects for life extension are likely to be shaped by those who feel strongly enough to do something about it, it is worth the rest of us thinking harder about what problem they think they are trying to solve. It is hard to see how it can be one of ultimate meaning, whatever the rhetoric suggests. Recommending a search for immortality seems to me a counsel of despair, not hope. As completely unlimited life is out of the question, what is the appeal of staking all on such a fantasy? If a life limited to 100 years is devoid of meaning, why would living to 200, or even 2000, improve matters? There would still be infinitely many years of non-being to follow. Of course, a much-extended life might enable one to survive into an era when human beings are modified in other ways, in which case the terms of the discussion will change in a manner we cannot predict. Then all bets are off. But until then, we seem to have an emerging lobby for the virtues of mere duration.

A contrasting view is offered by the environmentalist Bill McKibben, who suggests that we should agree on what is '*Enough*'.[6] McKibben assumes that most other people share his outlook on life. I think I might do. But I also see a great number of people who feel otherwise. Their spokesman is surely Ridley Scott's searingly memorable stand-in for any man meeting his maker in *Blade Runner* – the near-perfect but short-lived replicant, Roy, confronting Tyrell, the industrialist who supervised his design. Tyrell asks, nervously, what he can do for his creation. Roy cuts in harshly: 'I want more life, fucker.' Larkin would approve wholeheartedly.

Jon Turney is a science writer and editor, and teaches an MSc course in creative non-fiction at Imperial College London.

Notes

1 R Ettinger, *The Prospect of Immortality* (London: Sidgwick and Jackson, 1965).
2 A Harrington, *The Immortalist: An approach to the engineering of man's divinity* (London: Granada, 1969).
3 L Boia, *Forever Young: A cultural history of longevity* (London: Reaktion Books, 2004).
4 D Broderick, *The Last Mortal Generation* (Sydney, Australia: Reed Natural History, 1999).
5 J Hughes, *Citizen Cyborg: Why democratic societies must respond to the redesigned human of the future* (Boulder, CO: Westview Press, 2004).
6 B McKibben, *Enough: Staying human in an engineered age* (New York: Times Books, 2003).

10. Nip/Tuck nation

Decca Aitkenhead

The central London clinic of Transform Medical Group is a double-fronted Georgian town house just off Harley Street. In the waiting room, styled in the Mayfair fashion of reproduction furniture and gilt chandeliers, a dozen or so women in their 20s and 30s sit in silence, swivelling to scan each new arrival, but otherwise still and self-absorbed. The coffee-table glossies lie untouched; there is an air of quiet determination.

A widescreen television in the corner shows a corporate video, made in the manner of docu-reality TV, featuring emphatically ordinary patients with pronounced regional accents. A young man from the north-east phones home after an operation to pin back his ears, as overjoyed as a *Pop Idol* contestant, and a woman from the Midlands consults a doctor about her breasts. 'What kind of breasts would you like?' he asks. 'Where I didn't have to wear a bra,' she says. 'And they were just . . . there.'

Transform is the Burberry of the British cosmetic surgery industry, with its corporate mission to bring to the mass market the aesthetic of privilege. It has seen an annual growth rate of 10–20 per cent through the 1990s rocket recently to twice as much or more, and this is reflected across the industry as a whole. Cosmetic operations in BUPA hospitals were up by 32 per cent last year, male patient numbers more than doubled, and operations by the British Association of Aesthetic Plastic Surgeons (BAAPS) rose by 50 per cent. In the absence of a formal national record, the true number of cosmetic operations today is unknowable. BUPA puts it at around 75,000 a year, with another 50,000 non-surgical procedures such as Botox. By the end of last year the British market had been valued at more than £250 million.

Whatever the precise magnitude of the explosion, its impact on us has been overwhelming. A practice widely regarded not a decade ago as physically risky, morally doubtful, prohibitively expensive and socially embarrassing has been rebranded as something so innocuous and sensible as to be mundane. A survey this summer for *Grazia* magazine found that more than half of women now expect to have surgery. A quarter of teenage boys polled in May thought they might too, while more than 40 per cent of teenage girls said they had considered it. *Zoo Magazine* is currently running a competition for readers, in which the winner wins a breast augmentation for his girlfriend.

'It's like a big game we play, isn't it?' the editor of *Grazia* told me. 'What would you have done if you won the lottery? It's the thirty-something equivalent of the game you'd play at school, about who you would snog. We see beauty products and surgery as basically the same now.'

Acquitted of all its old political and psychological significance, cosmetic surgery has joined a humdrum spectrum of consumer lifestyle choice, alongside fashion and home furnishings. Radical transformations on this scale seldom, however, occur by accident. Who or what was responsible for changing our minds? And why were we so willing to be persuaded?

The media's fault?

I meet Transform's director of marketing in a consulting room. Liz Dale is instantly likeable, and the tone of our conversation is one I will come to recognise as characteristic of the industry – complicit and oddly intimate, occasionally giggly. She tells me that in the past six months there has been a dramatic rise in requests for cosmetic gynaecology, and we both squeal and widen our eyes, though it is hard to say whether we are intrigued or scandalised.

The pivotal turning point in public attitudes, Dale says, came five years ago when a number of British celebrities admitted to having Botox injections. 'It made it acceptable, because people didn't mind talking about Botox,' she says. 'And so, then, that naturally moved on to talking about having your nose corrected.' The talking was crucial, for although Transform advertises heavily through women's magazines and cable television, 'the single biggest source of our business by far comes from word of mouth.'

Everyone I speak to in the industry agrees that Botox is the key. Mel Braham, chairman of the Harley Medical Group, calls it the 'first foot in the door', but other factors are also identified. A buoyant economy and easy access to cheap credit have been important, as has advertising through the internet. As the market has expanded, economies of scale have made surgery cheaper in real terms, and tighter government regulations introduced this year have promoted an impression of greater safety. It is a happy collision of circumstances for the industry, but surely still not enough to account for its windfall.

Dale agrees, and then she slowly smiles. 'The media keep asking me this question, you know. They say, "What is it that's causing this explosion?" I'm sorry, but I have to say, "Well, it's you. It's the media."'

Cosmetic surgery stories are the media sensation this year. Market research has identified them as one of the single most effective incentives to make women buy a magazine, and the television schedules are dominated by makeover shows in which unhappy women undergo drastic head-to-toe surgery. In one American import, *The Swan*, the women then compete in a beauty pageant. MTV has had a big hit with *I Want a Famous Face*, which follows celebrity-obsessed patients on a dogged surgical quest to turn themselves into their icon. On Five, *Cosmetic Surgery Live* has brought us graphic footage of outlandish surgical desires from all over the world, including a man who paid to have his anus bleached.

'I defy you not to watch that moment when the curtain goes back, and the person sees what's happened to them,' says David Lyle, the producer of *The Swan*. 'It's almost the pornographic shot. Let's face it

– slapping a new coat of paint on is not as dramatic as having someone carve your face off.' Lyle is a cheerful, almost brash Australian who runs Reality Fox in America. He is very happy to talk about his work – but I soon discover he is the exception. When it comes to discussing what they do, there is a striking contrast in this country between those who actually provide surgery and those who make programmes about it. The former speak freely, even innocently, whereas the latter are amazingly cagey and defensive. There are endless press officers to be gone through, and questions to be emailed in advance, for all the world as though the media were not the messengers but the agents. And in a sense they are. When patients are asked what persuaded them to have surgery, with very few exceptions they cite what they have seen on television.

Does this make reality TV responsible for the growth of surgery? Richard Woolfe is the director of television for Living TV, which broadcasts *Extreme Makeover* and *The Swan*, and he represents the patrician school of reality TV theory. Woolfe believes his programmes provide a responsible public service, helping viewers navigate the modern world. 'We understand the needs and wants of our viewers, we talk to our viewers, and we take our responsibilities very seriously,' he says. 'After every plastic surgery programme we have a slot in the end board advising our viewers to contact their doctor before considering any procedure. . . . I think if people can learn from this experience, and take from it the good things, and do it properly and get proper advice, and realise there isn't a quick fix – that's a good thing.'

But there is something disingenuous in this public service argument. Living TV is not the NHS, nor even the BBC, but a commercial channel concerned with ratings. Lyle sounds more credible when he says, very briskly, 'So-called reality TV is, when all is said and done, entertainment. It doesn't make claims of any other nature, it doesn't set out to establish the social agenda. Our internal mission statement at Reality Fox is: "Hell no, we're not guilty!" We just think it's good fun – a guilty pleasure.'

There is one point on which Woolfe and Lyle do agree, however:

surgery will transform your life. 'The way I like to see it from a Living TV perspective,' says Woolfe, 'we are offering people the chance to go on an extraordinary journey. I feel proud about giving the chance to these people to change their lives completely. It's not about making people into the most beautiful specimens – this is about changing people's lives.'

But this consensus is also problematic, for it is the mantra of all surgeons that surgery cannot in fact transform your life. 'These programmes', the BAAPS remonstrates sternly, 'send a dangerous message to viewers, encouraging people to seek plastic surgery for the wrong reasons.' In other words, the single most powerful factor making people choose surgery is premised on a misconception. When I put this contradiction to the man at Five who makes *Cosmetic Surgery Live*, Ben Frow, he points out a trifle tartly that people like him are doing the surgeons' advertising for free.

Harmless fun?

It is hardly surprising that surgery has proliferated, when everyone involved is able to attribute agency elsewhere. Reality TV has the power to shape public opinion, without bearing any responsibility for the consequences. Surgeons profit from the popular belief that surgery can change your life without ever having to make the claim themselves. Both parties can reassure themselves that, like the makers of Botox or the lenders of cheap finance, they are merely doing their job – giving customers what they want. Because the real question is not which among them shares the greatest responsibility for surgery, but why so many of us want it.

From everything we read about surgery today, you might say we would be mad not to have it done. Patients are 'ecstatic' with the results, and beam into the camera for their 'after' photograph. The *News of the World* recently featured a 46-year-old woman who enthused after her facelift: 'I feel terrific! I look as good on the outside as I feel on the inside.' She had been blind for 12 years.

'It definitely works for people,' says Dale. 'We talk to patients afterwards, and they always say that it's given them confidence. We talk about confidence in our advertising, because that's the biggest thing they always say. You know,' she adds, 'it's nice to work in an industry that's happy. I see how happy people are when they have it done. It's like buying a new outfit – it makes you feel good.' It is said that staff entering the industry will have their first procedure within three months. 'I was here two weeks before they persuaded me to have Botox,' Dale notes with a smile.

There is more to cosmetic surgery than making patients happy, however. Surgical innovations can still provoke revulsion. Earlier this year on Channel 4, Richard and Judy featured a young woman who hated her feet so much that she found a surgeon in America to break her toes and trim off the ends. Richard was appalled, but Judy reasoned that if the feet had made the woman miserable, and fixing them made her happy, what was his problem? On these terms, any objection really is irrational – and so the boundaries of 'normal' get moved once again, and our instinctive misgivings are cast aside. But they are never really answered, only overruled, and traces of disquiet linger in the air.

It is interesting to see how we savage patients whose surgery goes wrong. When collagen injections disfigured Leslie Ash's lips, the hate campaign reached such a spiteful frenzy that the actress feared for her safety. Ash was bewildered, and it is easy to see why, for if surgery is as legitimate as we like to say, our only response could have been sympathy. Perhaps gleeful vengeance is the only acceptable way left to express a deep, unspeakable suspicion that something is not right.

For a large part of the twentieth century, patients who wanted cosmetic surgery would generally have been recommended therapy, their desires interpreted as an indication of pathology. Today this interpretation, if not quite eccentric, is rare. But what changed was not some major shift in psychological understanding, just the number of people now expressing the desire. Cosmetic surgery has not become popular because psychologists declared the desire 'normal'. Rather, the normality of surgery has been inferred from its popularity. It is, however, perfectly possible for an impulse to be both widespread and pathological. Last year, for example, 142,000 people were hospitalised in England and Wales from self-harm, the majority women, comfortably outnumbering cosmetic surgery in-patients.

Virginia L Blum has written one of the best books about surgery, having undergone two nose corrections herself as a young woman. In *Flesh Wounds*, she quotes a fellow patient, who says: 'I always looked in the mirror and thought, I want that bump out. I've thought, oh, I feel hideously ugly. But I've always thought, it's like you have a car that has a dent in it. If you got it fixed it would be quite a nice car. So I thought, apply the same thing to your nose.'

'Notice,' Blum writes, 'how her nose is both her and not her, something that makes her feel "hideously ugly" at the same time that it's as materially distinct as a car. This is what happens to your body when you start changing it surgically.' When cosmetic patients talk about their bodies, dissociation is a recurring theme, as though they no longer inhabit their own skin. At Transform, Dale tells me, women sometimes arrive to pay for their procedure with bags of cash. Plonking them down, they will joke, 'Look – here are my breasts!'

Reputable doctors are reluctant to operate on anyone who shows up with hundreds of celebrity photos, being wary of what they call 'unrealistic expectations', but the distinction between one and 100 photos seems less significant than the fact so many of us now need to look like somebody else. In the *Grazia* survey this summer, two-thirds of women under 25 said 'celebrities influence them into wanting surgery'. Whether or not they expect to look exactly like Angelina Jolie after surgery, her image has made their own intolerable.

By identifying with actresses and models and pop stars – people who really are judged on their looks – women exchange a three-dimensional identity for an image, and life becomes an unending audition, involving all the anxiety and rejection of *Pop Idol*. If you believe you have just ten seconds to make an impression, the only meaningful difference between makeup and surgery becomes price, and any amount will seem worth paying. But it is a poor exchange, for most women will never need to pass a Hollywood audition, and gain little from living every day as if they do.

There is, too, confusion around the discrepancy between patients' inner and outer account of themselves. They commonly complain that their external appearance is an impostor, obscuring the 'real' person they feel themselves to be. Toyah Wilcox decided to have a facelift because, 'I would look in the mirror and see someone tired, sad, grumpy, when inside I'd found my 40s the best time in my life. I was seeing someone that no longer represented me.' The singer was so thrilled with the results, she wrote a book about it. 'I now act differently, I'm different on stage, it has completely revolutionised my life. My self-esteem and my confidence are now my own. I'm not reliant on other people's opinion.'

Her before and after photographs are so distractingly astonishing that it is easy to miss the contradiction. But what Wilcox says is definitely odd. If her confidence and self-esteem were not 'her own' before surgery, it is hard to see how she could have been having 'the best time in her life', and easy to imagine why she would have felt grumpy. Far from misrepresenting her, the pre-surgery face may well have been telling the truth.

Many patients don't, of course, dispute the truth of their appearance. They just wish it were otherwise. Cosmetic surgery invokes the language of democratic freedom, granting universal access to the former privilege of beauty, and we like to speak of surgery as a choice. But what looks like choice from one angle can resemble coercion from another. Since the advent of cosmetic dentistry, the chairman of the Harley Medical Group observed to me, 'You look at someone and think, why on earth doesn't he get his teeth fixed up?' The people who come to him for surgery, he continues, 'just want to be normal. They don't want to stand out in the crowd, they just want to blend in.'

As I researched this article, I found myself studying my reflection differently, my features slowly looking less and less like me, and more like candidates for correction. It began feeling perverse to neglect the deep furrow in my brow, when I had been plucking my eyebrows for years. Why not try Botox? The results were amazing; I was delighted. But when I look in the mirror now, what I see are the laughter lines around my mouth, and I am wondering how much better I might look if I had them fixed too.

'In many ways', Blum observes, 'the wanting is partly the doing, inasmuch as you've already said yes to a whole host of surgery-related activities. . . . You have already pictured your surgically restructured body part.' In a postsurgical culture, she believes, the option of surgery has compromised all women's bodies, regardless of whether or not they take it. This may well be true – but it does not resolve the question many women find so difficult to answer. Do they owe it to themselves to look the best they can, or to each other to resist?

What about feminism?

Cosmetic surgery is an intractably feminist concern. More than 90 per cent of patients are female, and although the old feminist consensus against surgery has been dissolving since the 1990s, its special relationship to women is still taken for granted. Even those who find feminist discourse alienating have felt the need to draw an equation between surgery and empowerment, and they have done so very persuasively. Asked to name a role model they find inspiring, young women routinely cite Jordan.

The postfeminist case for surgery has been well put by Ann Robinson. As a 'champion of women', she considers her two facelifts proof of her sense of self-worth and self-respect – the independent choice of a liberated woman. When women tell Robinson their husband 'likes me as I am', she retorts: 'Of course he bloody likes you as you are. Safe in your box, unthreatening. As near to his mother as he can hope.' Women who won't have surgery 'lack self-belief'. A feminist listening 20 years ago would have found Robinson's inversion of female liberation bizarre. In a postfeminist age, it passes without notice.

Times have certainly changed, however. Whereas women a century ago were only expected to look like twentysomethings in their 20s, today they must keep it up for ever. 'Girls in their 30s who've had three children and suddenly they've got no breasts left?' says Norman Waterhouse, a surgeon and former president of BAAPS. 'That's a big problem, because women in their 30s these days, they're in the gym and the pool. They're out there. It doesn't make a 35-year-old feel desirable, it makes her feel like a washed-out old mum.' Waterhouse talks of his pleasure in restoring femininity – 'To give her back nice breasts, it's great' – and could see no viable alternative solution. 'People may say, "Oh, how superficial." But it's how it is, it's what I see.'

Feminism would once have expected to offer a viable alternative, but its unresolved attitude to beauty has created an ideological vacuum. Postfeminism was supposed to reclaim beautification as self-indulgence rather than man-pleasing, but it has generated a set of demands that are becoming limitless, leaving women no grounds for believing they have ever done enough.

In a book called *Reshaping the Female Body*, Kathy Davis tries to reconcile her traditional misgivings about surgery with the fact that even her friends are now considering it. She is uneasy with the old assumption that women must be 'cultural dopes' for succumbing. Equally, the old idea of blaming men holds little water when the majority of patients say their partners did not want them to have it. It should be possible to empathise with their decision, and still be angry about the circumstances that led them to it – one of which must be the legacy of postfeminism's love affair with beauty products.

What is the difference between highlighting your hair and having a facelift? As surgery gets safer and cheaper, women struggle to see how the latter could be bad, if the former is good. For feminism to offer a viable alternative to the surgical culture, it would have to risk reopening the argument about the entire continuum of the beauty industry. The exhausting regimes sold to women today as 'pampering' would need to be re-examined as a possible tyranny rather than a luxury. In a consumer culture, this is a daunting prospect. Having been branded hairy-legged militants a generation ago, it's not surprising that so few feminists seem willing to try.

I ask Waterhouse whether he could have predicted the explosion in his industry ten years ago. 'I don't think anyone could have,' he replies. Then, unprompted, he offers an instructive analogy. 'Would people have predicted *Big Brother* ten years ago – seeing people have sex live on TV?' To single out cosmetic surgery for special concern is, one might say, an arbitrary choice. If it is not the solution to anything, it is possibly not the problem either. For all the rhetoric of 'individual choice', surgery is a symptom of something much larger than the body – of faulty self-identity and celebrity obsession, and the transfer of moral authority from disinterested health professionals to the commercial media. Within the terms of a culture fashioned by shopping and cable TV, a facelift will probably always make sense.

I asked everybody I interviewed whether they could suggest anything that might slow or reverse surgery's growth. There were many blank faces. The media is unlikely to kill a golden ratings goose, after all, and the government is concerned only with tightening regulations, the effect of which is to increase consumer confidence. Feminists are too wary of sounding disloyal or unpragmatic to mount a coherent objection.

The only thing anyone could think of was a recession.

Decca Aitkenhead is a journalist and writer. This article first appeared in the Guardian *in September 2005.*

11. The perfect crime

Rachel Hurst

Man may be able to program his own cells long before he will be able to assess adequately the long-term consequences of such alterations, long before he will be able formulate goals, and long before he can resolve the ethical and moral problems which will be raised.

Marshal Nirenberg, Nobel Laureate, 1967

When we look for perfection in our children and ourselves, what are we looking for? Why are we looking for it? What sort of society do we want to create? Should society play around with natural selection and evolution? Where do human rights stand in all this? What role should those seen as imperfect have in decision-making? These questions should be the basis of our deliberations regarding genetic advances. Already some of these questions have been ignored in using some of the new techniques such as embryo selection, but that does not mean that it is too late to look at the issue as a whole. Ignoring these questions, searching for conformity and fearing diversity, could seriously damage our society.

What is perfection?

It would appear to be a universal perception that one of the ingredients of perfection and the pursuit of happiness is to be free from disabling impairment. This stigma and fear around disability has been recorded since ancient times, with tales of Spartans putting their babies out on the hillside to ensure survival of the fittest, and records of Romans committing infanticide and euthanasia. Religious doctrine has ruled disability to be a sign of sin and cultural myths and fears have confirmed disabled people as less than human.[1]

In 1883 Francis Galton proposed his theory of eugenics (derived from the Greek meaning 'good in birth') to improve the human stock of the nation. Although his theory was based on improving 'good' genes rather than eliminating 'bad' ones, his ideas were taken up enthusiastically as a basis for elimination. The Eugenics Education Society (later to become the Eugenics Society) was set up in Britain in 1907, and eugenics was widely supported by leading thinkers and politicians including George Bernard Shaw, John Maynard Keynes, Arthur James Balfour and Neville Chamberlain. They saw it as a moral and economic opportunity to eliminate depravity, destitution, vagrancy, criminality – as well as disabling impairment.[2] By 1926, similar societies were flourishing in the United States and legislation was enacted in the US, the UK and Scandinavia to compulsorily sterilise 'degenerates' – mostly disabled people. These eugenic practices were then made only too real and horrific by the Holocaust. It was disabled people who led the way to the gas chambers.

The Holocaust undoubtedly gave eugenics a bad name. Though it is interesting to note that the fact that disabled people, gypsies and homosexuals were rounded up and gassed (in the case of disabled people, long before Jews) was not reported as part of the Holocaust until fairly recently. The prewar eugenic world with its compulsory sterilisation and isolation of disabled people into separate institutions would have had some sympathy with these deaths. It is only the rights movements of these groups that have since highlighted the Nazi attempt at their elimination.

An important ethical question is: whose idea of perfection is the right one? Peter Singer, the professor of ethics who believes that there is more value in the life of an intelligent baboon than of a severely disabled child, gave a brief outline of himself in reply to this question. No doubt he was being semi-humourous, but parents, either secretly or overtly, do want their children to be like the best bits of themselves. And how many parents are there who find it hard to accept a child for what they are rather than what they wanted?

Why are we looking for perfection?

> *Soon it will be a sin for parents to have a child which carries the heavy burden of genetic disease. We are entering a world where we have to consider the quality of our children.*
>
> Dr Bob Edwards

Many modern ethicists and geneticists do not see the pursuit of healthy and 'perfect' babies as mirroring the state-endorsed laws and practices of Galton eugenics. They believe that the possibilities of the new molecular genetics are beneficial, preventing suffering and providing positive benefits both to individuals and society.

There is little doubt that the modern world is not a good place for disabled people to be born into. Although medical science has come a long way and far more disabled people are surviving accident and trauma, services and support are very poor or non-existent. Sixty per cent of disabled people in the UK live in poverty, isolated in inaccessible homes from an inaccessible world. Attitudes to disabled people are extremely negative and expectations of their participation, integrity and capacity severely limited. Research data has shown that 10 per cent of violations of rights against disabled people have been denial of the right to life – through murder, neglect or withdrawal of treatment. A further 43 per cent of violations in the UK are of degrading and inhuman treatment.[3] Several murders by family members are considered by the courts as 'mercy killings' and the perpetrators given non-custodial sentences for the lesser charge of manslaughter.

Many disabled people themselves, as well as the general public, do not comprehend that it is these negative attitudes to disabled people's quality of life, fears of supposed suffering and the barriers to participation and equality that are the disabling factor – not the impairments. As a mother of a young girl with Down's syndrome says:

> *Margaret does not view her life as unremitting human suffering (although she is angry that I haven't bought her an iPod). She's consumed with more important things, like the performance of the Boston Red Sox in the playoffs and the dance she is going to this weekend. Oh sure, she wishes she could learn faster and had better math skills. So do I. But it does not ruin our day, much less our lives. It's the negative social attitudes that cause us to suffer.*[4]

The cultural and political ideologies underpinning the new genetics work to a medical model of disability, seeing disabled people as solely consisting of their impairments – not their intrinsic humanity. This medicalisation of disability not only leads to discrimination and underpins eugenic practices but also justifies massive expenditure on research into genetics. By contrast, if a social model of disability[5] were adopted it would be the negative barriers of the environment that would be the subjects of research and financial commitment. And many of the costs of ensuring disabled people have a decent life could then be offset by the benefits of inclusion for both them and their families (who are generally dragged into poverty with their disabled member).

Modern culture, aided and abetted by the media, fosters a climate of physical perfection. We are all supposed to look like models and live highly social and active lives. Worship at the altar of shopping has overtaken religion or the enjoyment of life itself. This pursuit of perfection has produced much distress and has, in some cases, led to mental breakdown, eating disorders and body mutilation.

We are not used to suffering and do everything in our power to deny that it happens to us. And yet this very denial must be one of the causes of the rise in work-related stress and post-traumatic stress disorders. We do not talk about death or allow mourners the space and time to grieve. We react by giving money to alleviate other people's suffering from the comfort of our own homes, but are afraid to give our own time and support to our neighbours for fear of aggressive or unrewarding responses.

Envisioning alternatives

In view of all these negative attitudes, it is hardly surprising that people want perfection. It seems the answer. But is it? Surely we have learnt from history that diversity in society is essential for its health and survival. Even a rudimentary knowledge of evolution and natural selection has shown that if a species does not learn to adjust to differing environmental impacts, it does not survive. We need a diverse society for the survival of all.

Striving for perfection is itself a burden for most people and an impossible and unaffordable goal. Instead, how much better to put our energies into celebrating diversity and our money into making diversity and equality work. Margaret's mother again:

> *Margaret is a person and a member of our family. She has my husband's eyes, my hair and my mother-in-law's sense of humour. We love and admire her because of who she is – feisty and zesty and full of life – not in spite of it. She enriches our lives. If we might not have chosen to welcome her into our family, given the choice, then that is a statement more about our ignorance that about her inherent worth.*
>
> *What I don't understand is how we as a society can tacitly write off a whole group of people as having no value. I'd like to think that it's time to put that particular piece of baggage on the table and talk about it, but I'm not optimistic. People want what they want: a perfect baby, a perfect life. To which I say: Good luck. Or maybe, dream on.*[6]

The world's response to the horrors of the Holocaust and the Second World War was to write the Universal Declaration of Human Rights. At that time, the global community, reeling from the atrocities, agreed that every human being, without discrimination of any kind, was born with the same right to equality, freedom and dignity. Inevitably, with such encompassing and high ideals, no state has implemented these rights for all its citizens, but the majority of societies work hard to fulfil their obligations under the various human rights treaties.

Yet no society has gone even halfway to ensuring rights for its disabled citizens. There are, around the world, many legislative procedures that deny disabled people equality in justice and in the right to life. In some countries, disabled people are not allowed to own property, bank accounts, access justice, vote, marry or have children.[7] Abortion laws say that it is legal to abort a viable fetus if it has a disabling impairment, embryos are routinely eliminated on the grounds of impairment (though it is illegal on the grounds of race or gender) and, as already noted, murder of a disabled person can be seen as 'merciful' and go unpunished.

We will never be able to continue building a society based on human rights while genetic advances are directed towards the elimination of disabling impairment. The most important right – the right to life itself – can never be ensured in this climate.

Further problems arise from legislation supporting assisted suicide or voluntary euthanasia, and from the pursuit, in the courts, of compensation for a 'wrongful life' – a life that is deemed wrongful because doctors have allowed the birth of a child that they should have suspected would be disabled. With both these issues it is the lack of services – of financial and personal support and health care – that drives individuals to seek either death or compensation. Talk of relieving pain and suffering hides the unspoken concerns of the so-called costs of a disabled life.

The contribution of disabled people

What role should any group play in the political and ethical dialogue about their elimination? The answer to that question would be straightforward if we were talking about other groups who face discrimination and injustice. If we were talking about women or ethnic minorities the answer would be easy: their involvement in decision-making is essential, and opportunities to develop their voice must be given real priority. But for disabled people the response has been different. For a long time, medical and rehabilitation professionals or charitable service providers have been seen as representing the voice of disabled people. In the realm of genetics, it is the ethical and genetic commissions and agencies, mainly consisting of non-disabled professionals and academics, who take the lead in the decision-making.

How can disabled people make informed input into these discussions? If you live, as most disabled people do, in a world that tells you that you are less than human, that your life is not worth living, that you cost too much, what does that do to your self-worth? If what you give to your family and friends – even if it is only the gift of love – is seen as worthless, how do you not react as society wants you to and agree that people like you should be eliminated? If you are not given the tools and support to participate, if you do not have access to proper health care and pain relief, why should you want to go on living?

It is only when disabled individuals have understood the true meaning of disability – that it is not their personal fault or problem, but is created and perpetuated by a disabling environment – that they can be liberated into recognising their right to dignity, equality and self-respect. Only then will they feel able to fight against these negative social responses.

Some ethicists, who are in favour of pursuing perfection, do hint that they would include themselves in the elimination of faulty genes. But that is easy to say when you are safely alive and nothing is actually being done to get rid of you. For disabled people, the constant reminder that some people think that people like you would be better off dead is a sobering, stressful experience and a real threat to your intrinsic humanity.

Rachel Hurst is Director of Disability Awareness in Action.

Notes

1 A Fletcher, *Obstacles to Overcoming the Integration of Disabled People* (London: Disability Awareness in Action, 1995).
2 A Kerr and T Shakespeare, *Genetic Politics* (Cheltenham: New Clarion Press, 2002).
3 R Light, *A DAA HR Data Summary Report* (London: DAA, 2004).
4 PE Bauer, 'The abortion debate no one wants to have: prenatal testing is

making your right to abort a disabled child more like "your duty" to abort a disabled child', *Washington Post*, 18 Oct 2005. See: www.washingtonpost.com/wp-dyn/content/article/2005/10/17/AR2005101701311_pf.html (accessed 4 Jan 2006).

5 World Health Organization, *International Classification of Functioning, Disability and Health* (Geneva: WHO, 2003). This classification defines disability as the interaction between people with impairments and negative environmental barriers, including those of attitude and belief.

6 Bauer, 'The abortion debate'.

7 T Dube et al, *Promoting Inclusion? Disabled people, legislation & public policy*, a research report for Knowledge & Research Project, Theme 22 (London: Department for International Development, 2005).

12. The unenhanced underclass

Gregor Wolbring

Advances in the converging technological fields of nanotechnology, biotechnology, information technology and the cognitive sciences are set to increase our abilities to enhance our bodies and brains in terms of structure, function or capabilities. They will do this beyond the typical boundaries of what it means to be human to the point where the technical description of us as members of the species *Homo sapiens* ceases to be accurate. Many different forms of enhancement are proposed with many different purposes. Each form and purpose of enhancement comes with its own sales pitches, social consequences, problems and implications.

One of the main arguments in the enhancement debate is that you can and should make a distinction between therapy and enhancement. However, this argument and many others employed in the enhancement debate depend on what concept of health you follow. So for me the key question is which concepts of health, disease, disability, well-being and even medicine we use. I'd also like to highlight a number of dynamics that make it nearly impossible to prevent enhancements and some of the problems and policy implications this could cause.

Models and causes of health and well-being

First of all we need to clarify the difference between 'health' and 'well-being'. The World Health Organization (WHO) considers well-being as being within the umbrella term 'health' where health is defined as 'a state of complete physical, mental and social well-being and not merely the absence of disease or infirmity'.[1] This model combines medical health and social health under the term health. But, increasingly, the policy world is moving away from the WHO definition of health and treating well-being less and less as a determinant of health. Policy-makers are interpreting the term health to mean medical health or medical illness. 'Social health' is often not covered under this definition of health.

Second, we need to look at the existing main models of health and disease – the medical and social models. Within the medical model of health and disease, health is limited to cover 'medical health' and is characterised as the normative functioning of biological systems, whereas disease or illness is defined as the sub-normative functioning of biological systems. This model does not deal with social well-being or 'social health'. Its method for locating the cause of and solution for 'ill medical health' comes in two flavours:

- identifying the cause of sub-normative functioning within the individual's biological system leading to medical interventions that bring the individual back towards the species typical norm (these are medical, individualistic cures)[2]
- external factors such as contaminated water, which leads to bacterial or parasitic infections, or job insecurity, which contributes to stress and heart disease.

If people refer to and talk about the 'social model of health' or the 'social determinants of health', they are mostly talking about the social causes of medical health, looking at how social factors contribute to medical illness. However, the real social model of health does not cover just social causes of medical health, but also the social well-being – the 'social health' – of a person who is not medically ill. One can be in bad social health without having to be in bad medical health. Under the social model, disabled people are not disabled by their impairment but by society's inability to adapt to them.

The transhumanist model

But now there's a new kid on the block to add to the two previous models. Within the transhumanist or enhancement model, health is no longer characterised as an endpoint, where someone is healthy if their biological system functions within the normal boundaries. No matter how conventionally 'medically healthy' a person is, a person is seen as limited and defective, in need of constant improvement made possible by new technologies appearing on the horizon. Think of it as a little bit like the constant software upgrades we do on our computer. Health, in this model, is the concept of having obtained maximum enhancement of one's abilities, functioning and body structure. Disease, in this case, is identified through a negative self-perception of one's unenhanced body or a negative perception of social groups who are confined to unenhanced human bodies.

Under this model, technologies which add new abilities to the human body are seen as the remedy for ill health and well-being. Enhancement medicine is the new field providing the remedy through surgery, pharmaceuticals, implants and other means.

To see the differences between the three models of health, and in particular the potential effect of the transhumanist model, look at these quotes from the Bangkok Charter for Health Promotion in a Globalized World[3] and the Universal Declaration of Human Rights[4] and think about what using the different definitions of the terms health and well-being above would mean for the scope and actions required:

> *The United Nations recognizes that the enjoyment of the highest attainable standard of health is one of the fundamental rights of every human being without discrimination.*

> *Regulate and legislate to ensure a high level of protection from harm and enable equal opportunity for health and well being for all people.*

Government and international bodies must act to close the gap in health between rich and poor.

Everyone has the right to a standard of living adequate for the health and well-being of himself.

The consequences of enhancement

I don't believe that we can prevent human enhancement technologies from developing. This poses some imminent problems that current systems of governance for science and technology are unlikely to be able to deal with. Here are a few of the problems I foresee.

First, the question of personhood. All UN-based documents use the term 'person'. However, the term is not set in stone. Throughout history, many humans have not been seen as persons and in some places some are still seen as non-persons today. How do we define human beings? What happens when we go beyond what can be defined scientifically as *Homo sapiens*? What are the criteria for personhood? Do we have to redefine personhood to take into account new technological realities? How does any given redefinition of personhood affect people perceived as persons today? Might some people who are perceived as persons today become non-persons? These are all questions that human enhancement raises.

Second, the creation of an ability divide. The more forms of enhancement become available, the bigger the ability divide will become. This would follow the pattern of the divides that developed after the introduction of other technologies. As we seem not to be able to close any of the other divides (remember 98 per cent of webpages are still not accessible to blind people), it is doubtful we will be able – under current policies – to close the ability divide. Indeed, people and groups who promote human enhancement use the existence of other societally accepted divides to further their cause. As the World Transhumanist Association states: 'Rich parents send their kids to better schools and provide them with resources such as personal connections and information technology that may not be available to the less privileged. Such advantages lead to greater earnings later in life and serve to increase social inequalities.'[5] A debate has to take place about which divides are acceptable, under what conditions, and why.

A third, related, problem will be a worsening of the gap between the rich and the poor. Transhumanists and others propose that wealth will eventually trickle down.[6] However, if this is the case, why do we still have poor people, unclean water and many places without phones and electricity? Every technology has led to a new group of marginalised people and to new inequalities. There is no reason under today's policy realities why this would be different if the human body becomes the newest frontier of commodification. As much as human enhancement technology will become an enabling technology for the few, it will become a disabling technology for the many. I believe that we need to change the whole system towards distributive justice, giving the enhancements first to the ones who need them most. And as this is not very likely to happen, the second best option is to ensure absolutely that no one can gain any positional advantage from enhancements and no one can force their desires and self-perception on others, whether it is their child or child to be or others. If we go on as we are today we will see the appearance of a new underclass of people – the unenhanced.

Fourth, it will require changes to the concept of responsibility. The transhumanists consider it to be a parental responsibility to use genetic screening and therapeutic enhancements to ensure as 'healthy' a child as possible.[7] Under such a model, would it be child abuse if parents refused to give their children cochlear implants, if they felt there was nothing wrong with their child using sign language, lip reading or other alternative modes of communication? Would it be child abuse to fail to provide a 'normal' child early in life with a brain–machine interface?

Finally, it will increase the number of people perceived as 'impaired' because as enhancement technologies are developed, those defined as 'impaired people' will change. The transhumanist model sees every human body as defective and in need of improvement, such that every unenhanced human being is, by definition, 'disabled' in the impairment or medical sense. For the transhumanists, disabled people are those who are not able to improve themselves beyond what is normal for our species (I call these people the techno-poor disabled). Theirs is a variation of the medical, individualistic model using transhumanist principles, taking the medical model further to include enhancement technologies.

It might be assumed that 'traditional disabled people' would welcome such a shift, as it would move the focus away from particular forms of impairment, towards the ability to enhance oneself – a challenge that the 'traditional disabled people' would share with other 'unenhanced people'. Indeed, many transhumanists are very aware of the potential to use disabled people as a trailblazer for the acceptance of transhumanist ideas and products.[8] As James Hughes, the executive director of the World Transhumanist Association, writes, 'Although few disabled people and transhumanists realise it yet, we are allies in fighting for technological empowerment.'[9]

However, as many 'traditional disabled people' are poor and live in low income countries they have far more to lose than gain from such a shift. They might think that they are better off because they would share that lack of ability with others who can't afford the enhancement, but we can expect that resources would never be 'wasted' on people who are below the traditional norm. This is because with the same amount of money more people who already fit the traditional norm could be enhanced than people who are different.

As Murray and Acharya have written (Murray is the father of 'disability adjusted life years' – a measure developed to give decision-makers a tool to judge who money should go to in health interventions), 'individuals prefer, after appropriate deliberation, to extend the life of healthy individuals rather than those in a health state worse than perfect health'.[10] What this means is that it is realistic to expect that if we follow the same model decision-makers will choose to enhance the lives of healthy individuals rather than those in a state of less than perfect health because it will be seen as better value for money.

What all these problems combine to mean is that, unless we act now, we are sleepwalking into a society with an unenhanced underclass.

Dr Gregor Wolbring is a biochemist, bioethicist, health researcher, futurist and disability studies and governance of science and technology scholar with appointments at a number of universities. His webpage is www.bioethicsanddisability.org/start.html

Notes

1 World Health Organization. WHO definition of health: preamble to the Constitution of the World Health Organization as adopted by the International Health Conference, New York,19–22 June1946; signed on 22 July 1946 by the representatives of 61 states (*Official Records of the World Health Organization*, no. 2, p. 100) and entered into force on 7 April 1948. The definition has not been amended since 1948. Available at: www.who.int/about/definition/en/ (accessed 11 Jan 2006).

2 G Wolbring, 'Solutions follow perceptions: NBIC and the concept of health, medicine, disability and disease', *Alberta Health Law Review* 12, no 3 (2004).

3 The 6th Global Conference on Health Promotion, 'Policy and partnership for action: addressing the determinants of health', Bangkok, Thailand, 7–11 Aug 2005, available from: www.who.int/healthpromotion/conferences/6gchp/en/ (accessed 11 Jan 2006); see also: 'The Bangkok Charter for Health Promotion in a Globalized World', available from www.who.int/healthpromotion/conferences/6gchp/bangkok_charter/en/print.html (accessed 11 Jan 2006).

4 See: www.un.org/Overview/rights.html (accessed 11 Jan 2006).

5 See: www.transhumanism.org/resources/faq.html#31 (accessed 11 Jan 2006).

6 Ibid.

7 See: www.transhumanism.org/resources/faq.html (accessed 11 Jan 2006).

8 See: http://transhumanism.org/index.php/WTA/communities/physicallydisabled/ (accessed 11 Jan 2006).

9 J Hughes, 'Battle plan to be more than well: transhumanism is finally getting in gear', 3 Jun 2004; available at: www.betterhumans.com/Features/Columns/Change_Surfing/column.aspx?articleID=2004-06-03-1 (accessed 11 Jan 2006).

10 CJ Murray and AK Acharya, 'Understanding DALYs (disability-adjusted life years)', *Journal of Health Economics* 16, no 6 (1997).

13. Does smarter mean happier?

Raj Persaud

We live in a competitive world, and one that's likely to stay so for the foreseeable future, as free markets appear to be the most efficient and effective way of organising ourselves. Enhancement technology will be used to make us better at things where there is already a sense of competition, and we will strive most to enhance ourselves in the most competitive areas of life.

This means we are more likely to focus on making ourselves smarter (because there are exams to be passed and promotions to be garnered) than on prioritising making ourselves more kind. There seems little competition for kindness or obvious reward for it (at least in a free market).

This skew in thinking about enhancement means, bluntly, that helping ourselves be smarter will be the priority – but are there downsides to being cleverer which might be overlooked in the race for elevated brain power? Also, given that new technology rarely benefits everyone equally, what are the problems inherent in a society where greater inequality in cleverness beckons for particular groups who can afford new enhancement technologies?

Smartness and suicide

These questions may appear abstruse to those outside the field of psychology and psychiatry, but within these disciplines real questions have been raised about the possibility that smart people are more prone to suicide. If that is the case, seeking to raise IQ generally, regardless of the social or emotional consequences, would be cavalier.

The issue of whether having a higher IQ raises your chances of suicide is an intriguing one that psychologists have been grappling with for many years. A definitive answer may have been provided recently by one of the biggest studies conducted in the area by Martin Voracek, an academic at the University of Vienna Medical School, Austria. Voracek, in a study published recently in the prestigious journal *Personality and Individual Differences*,[1] compared suicide rates in 85 countries across the world with intelligence levels. The curious result is that the higher the average IQ in a country, the higher the suicide rate. The association is extremely statistically significant.

Voracek got the idea for his survey from the long-recognised fact that suicide rates are higher among college students than for same-age but less-educated young adults. One study even found that those at university who kill themselves tended to have had above-average grades compared with the general student body. Voracek argues that perhaps the strongest evidence that those with very high IQs are more likely to kill themselves comes from the 'Terman Genetic Study of Genius'. For this unique study, 1528 gifted children (857 boys and 671 girls), who were on average 11 years old, were identified in Californian public schools during the 1921/22 school year and were followed up over their entire life cycle. The inclusion criterion was an IQ of 140 or higher, meaning that all study participants ranked within the top 1 per cent of the population for intelligence. The average IQ of the group was 151.

During the observation period, 34 participants committed suicide, a rate almost three times that of the enrolment site for the study – California – and roughly four times the suicide mortality for the general population. But Voracek points out that the high suicide mortality in the Terman study is even more interesting in light of the fact that this sample was also found generally to live much longer and be in better physical health, relative to the corresponding general population. As a consequence, one in 11 male deaths and one in 19 female deaths in the Terman sample were from suicide, which is an extremely skewed cause for mortality compared with the general population.

The burden of being brainy

If the accumulating evidence points to a higher suicide rate among the intellectually talented why might that be? Those with higher IQs tend to be more successful in life generally, across several domains, than those who are less smart. Indeed, Charles Murray, co-author of the best-selling book on IQ, *The Bell Curve*,[2] which controversially argued that there were profound racial and gender differences in IQ, points out that as society becomes more technological, IQ will increasingly determine success in life. Being smart is going to matter more and more in the future, whereas in the past being physically strong or being born to powerful or rich parents was the key. As technological advances ensure that complexity, speed and change will increasingly be the key features of our society, we will all need to be smarter to contend with these changes.

If this is the case, why then the higher suicide rates among smarter people? One theory is that perhaps the unusual amount of self-awareness and desire to excel among the brainy means they put more pressure on themselves. Yet surely the clever should have the insight to see what they are doing to themselves?

Robert Sternberg, Professor of Psychology at Yale University and one of the world's leading authorities on IQ, isn't so sure. He argues that smart people can act foolishly by virtue of thinking they are too smart to do so. He points out some key cognitive fallacies that those with higher IQs are paradoxically more prone to which result in foolish behaviour.

The first is 'unrealistic optimism', whereby Sternberg argues the clever believe they are so clever that they can do whatever they want and not have to worry about it. Another feature of the brainy, according to Sternberg, is 'egocentrism', whereby they focus on themselves and what benefits them while discounting or even totally ignoring their responsibilities to others, who are less smart. A particularly grave cognitive error the intellectual commit is that of 'omniscience', whereby they believe they know everything, instead of realising that they don't know everything.

Sternberg proposes that the key to life is not to strive to be clever, but rather to be wise. This requires, he contends, an interest in the common good as a way of surviving in a world full of less intelligent, but no less worthy, people. The curious implication of Sternberg's thesis is that the smart may find it tough living in a non-smart world, and need therefore to be aware of how to get on with those with lower IQs as a key life skill. Ironically, then, highly intelligent individuals may on average be less adapted to general living contexts, and as a result could be more prone to alienation from others, and therefore to suicide.

So it seems once you realise you are smart, the next key challenge is to find a way of getting on with those around you who don't share your IQ, without dumbing down to an extent that depresses you. This could explain why those with high IQs cling on to each other for dear life – literally – when they find each other.

Enhancing emotional intelligence

Given the problems inherent in raising IQ or producing more IQ equality in society, perhaps it might be better to focus more on raising 'EQ' or emotional intelligence. Emotional intelligence was a concept introduced to explain why so many with obviously high IQs do not seem to advance as expected in the real world, away from academia or the IQ testing station. The notion is that some people are brainy when it comes to pencil and paper tests but don't know how to get on with others. They lack social skills, and this explains why they can checkmate you in four moves but can't necessarily work out what to say or wear at an interview.

Having a high EQ means you can manage your emotions, and better recognise and influence the emotional state of others. Empathy and persuasion are key characteristics of those with high EQ, so maybe we should be trying to enhance EQ rather than IQ, as EQ helps us to cope better with differences in each other.

There is already a group of people among us who appear to have superior skills in this arena. Yet if these people are currently sidelined, what hope do we have of producing higher EQ generally in society at large? Their relative marginalisation or lack of success suggests EQ is not valued in the way it needs to be for it to be part of a human enhancement programme. This group of people are technically referred to within the field as women.

Gender, conflict and happiness

An example of the benefits of being a woman when it comes to reducing conflict was effectively demonstrated by some intriguing research conducted by two political scientists in the United States (one of whom was a man incidentally). The researchers looked at all the countries involved in international conflicts around the world over the last 50 years and found that the more women were involved in the leadership of a society, the less militarily aggressive the society was, and the lower the probability of violent conflict with other countries.

The researchers, Mary Caprioli and Mark Boyer, argue that their study, which was published in the *Journal of Conflict Resolution*,[3] is strong evidence for the proposition that, generally, women work for peace and men wage war. Women are more likely to use a collective or consensual approach to problem-solving, rather than an approach that focuses on the unilateral imposition of solutions.

Psychologically, at quite a profound level, men tend to engage in power struggles for personal gain, whereas females tend to attempt to minimise power differences, to share resources, and to treat others equally. Yet despite these advantages of female leadership, according to the research conducted by Caprioli and Boyer, only 24 countries around the world have placed a female leader in office since 1900. Only 16.6 per cent of these countries led by a woman were involved in international crises at any point during the period of female leadership, and none of these female leaders initiated the crises.

The researchers used political equality, measured as the percentage of women in parliament, as a measure of gender equality within society. Put simply, their finding is that as the percentage of women in the legislature of a country increases, the less severe is the violence between countries. Indeed, if the percentage of women in the legislature increases by 5 per cent, a state is nearly five times less likely to use violence internationally.

In terms of the current warlike position of the USA compared with more pacifist Europe, it is interesting to note that the US has far fewer women in its legislature compared with most European countries – for the US the figure is just over 14 per cent compared with Sweden at 42 per cent. Indeed, Scandinavian countries take the top six consecutive spots in the world league table for highest female representation in parliament – followed by Germany with 32 per cent. The UK, which has arguably been more aggressive in recent conflicts than the rest of Europe, is down at 17.9 per cent.

One theory behind this, argue Caprioli and Boyer, is that competition, violence, intransigence and territoriality are all associated with a male approach to international relations. Women, on the other hand, are less likely to see crisis negotiation as a competition or to advocate the use of violence as a solution. That said, female leaders are often perceived to be just as aggressive as men. Leaders of recent years such as Margaret Thatcher, Benazir Bhutto, Indira Ghandi and Golda Meir were seen as hawks rather than doves, and all were caught up in violent conflicts.

But perhaps female leaders must also contend with negative perceptions from male opponents. For example, gender was a factor in the events and resolution of the 1971 Indo–Pakistan war in which Indira Ghandi had a key role. Caprioli and Boyer remind us that President Yahya Khan of Pakistan stated that he would have reacted less violently and been less rigid as the leader of Pakistan in the conflict with India if a male had headed the Indian government. Indeed, President Khan was quoted as saying: 'If that woman [Indira Gandhi] thinks she is going to cow me down, I refuse to take it.' So the behaviour of male leaders when faced with a female opponent becomes a factor – a sense of macho pride which makes them unwilling to 'lose' to a woman, lest their masculinity be questioned.

Female leaders who have risen to power through a male-dominated political environment may well need to be more aggressive than their male counterparts in crisis, argue Caprioli and Boyer. Although differences exist in male and female leadership styles, women in positions of power may find themselves compelled to convey their strength in traditional male terms. And they may also work harder to 'win' in a crisis for the same reasons, because to respond in a more feminine way would be seen as 'weakness' and would be political suicide.

Caprioli and Boyer's research suggests that we don't just need more women in parliaments and legislatures, but also to live in societies that embrace more feminine values, so that women who succeed will feel less pressure to be more like men. This view is supported by Ruut Veenhoven of Erasmus University, a leading expert on happiness who recently published a study that found all over the world people are happier in more feminine nations.

Veenhoven defines masculine cultures as those which expect men to be assertive, ambitious and competitive, to strive for material success, and to respect whatever is big, strong and fast. These cultures expect women to serve and to care for the non-material side of life, for children and the weak. Feminine cultures, on the other hand, define relatively overlapping social roles for the sexes, in which men need not be ambitious or competitive but may go for a different goal in life than material success; men may respect that which is small, weak and slow.

So, in more masculine cultures (such as Japan, Austria and Venezuela) political and organisational values emphasise material success and assertiveness, whereas in more feminine cultures (like Sweden, Norway and the Netherlands) they accentuate other values, interpersonal relationships, and sympathy and concern for the weak.

If people are happier in feminine societies and these countries tend to get involved in less conflict with their neighbours, maybe the key enhancement that will produce most well-being in the future would be for us to become in some senses more feminine. This, in the sense conveyed by this research, means more empathic, kind and caring, more aware of others' emotional states and more able to influence our own and others' emotions.

This should be our priority rather than merely aiming to raise IQ. Ironically, this enhancement strategy requires us not to become different in order to improve, but rather to become more like the good parts of ourselves. Enhanced people are already walking around among us, but we tend to ignore them. We do this at our peril and new technologies will not save us from this mistake.

Raj Persaud is Consultant Psychiatrist at the Bethlem Royal and Maudsley Hospitals and Gresham Professor for Public Understanding of Psychiatry.

Notes

1 M Voracek, 'National intelligence and suicide rate: an ecological study of 85 countries', *Personality and Individual Differences* 37 (2004).
2 RJ Herrnstein and C Murray, *The Bell Curve: Intelligence and class structure in American life* (New York: Simon and Schuster, 1996).
3 M Caprioli and M Boyer, 'Gender, violence, and international crisis', *Journal of Conflict Resolution* 45, no 4 (2001).

DEMOS – Licence to Publish

compensation. The exchange of the Work for other copyrighted works by means of digital file-sharing or otherwise shall not be considered to be intended for or directed toward commercial advantage or private monetary compensation, provided there is no payment of any monetary compensation in connection with the exchange of copyrighted works.

- **c** If you distribute, publicly display, publicly perform, or publicly digitally perform the Work or any Collective Works, You must keep intact all copyright notices for the Work and give the Original Author credit reasonable to the medium or means You are utilizing by conveying the name (or pseudonym if applicable) of the Original Author if supplied; the title of the Work if supplied. Such credit may be implemented in any reasonable manner; provided, however, that in the case of a Collective Work, at a minimum such credit will appear where any other comparable authorship credit appears and in a manner at least as prominent as such other comparable authorship credit.

5. Representations, Warranties and Disclaimer

- **a** By offering the Work for public release under this Licence, Licensor represents and warrants that, to the best of Licensor's knowledge after reasonable inquiry:
  - **i** Licensor has secured all rights in the Work necessary to grant the licence rights hereunder and to permit the lawful exercise of the rights granted hereunder without You having any obligation to pay any royalties, compulsory licence fees, residuals or any other payments;
  - **ii** The Work does not infringe the copyright, trademark, publicity rights, common law rights or any other right of any third party or constitute defamation, invasion of privacy or other tortious injury to any third party.
- **b** EXCEPT AS EXPRESSLY STATED IN THIS LICENCE OR OTHERWISE AGREED IN WRITING OR REQUIRED BY APPLICABLE LAW, THE WORK IS LICENCED ON AN "AS IS" BASIS, WITHOUT WARRANTIES OF ANY KIND, EITHER EXPRESS OR IMPLIED INCLUDING, WITHOUT LIMITATION, ANY WARRANTIES REGARDING THE CONTENTS OR ACCURACY OF THE WORK.

6. Limitation on Liability. EXCEPT TO THE EXTENT REQUIRED BY APPLICABLE LAW, AND EXCEPT FOR DAMAGES ARISING FROM LIABILITY TO A THIRD PARTY RESULTING FROM BREACH OF THE WARRANTIES IN SECTION 5, IN NO EVENT WILL LICENSOR BE LIABLE TO YOU ON ANY LEGAL THEORY FOR ANY SPECIAL, INCIDENTAL, CONSEQUENTIAL, PUNITIVE OR EXEMPLARY DAMAGES ARISING OUT OF THIS LICENCE OR THE USE OF THE WORK, EVEN IF LICENSOR HAS BEEN ADVISED OF THE POSSIBILITY OF SUCH DAMAGES.

7. Termination

- **a** This Licence and the rights granted hereunder will terminate automatically upon any breach by You of the terms of this Licence. Individuals or entities who have received Collective Works from You under this Licence, however, will not have their licences terminated provided such individuals or entities remain in full compliance with those licences. Sections 1, 2, 5, 6, 7, and 8 will survive any termination of this Licence.
- **b** Subject to the above terms and conditions, the licence granted here is perpetual (for the duration of the applicable copyright in the Work). Notwithstanding the above, Licensor reserves the right to release the Work under different licence terms or to stop distributing the Work at any time; provided, however that any such election will not serve to withdraw this Licence (or any other licence that has been, or is required to be, granted under the terms of this Licence), and this Licence will continue in full force and effect unless terminated as stated above.

8. Miscellaneous

- **a** Each time You distribute or publicly digitally perform the Work or a Collective Work, DEMOS offers to the recipient a licence to the Work on the same terms and conditions as the licence granted to You under this Licence.
- **b** If any provision of this Licence is invalid or unenforceable under applicable law, it shall not affect the validity or enforceability of the remainder of the terms of this Licence, and without further action by the parties to this agreement, such provision shall be reformed to the minimum extent necessary to make such provision valid and enforceable.
- **c** No term or provision of this Licence shall be deemed waived and no breach consented to unless such waiver or consent shall be in writing and signed by the party to be charged with such waiver or consent.
- **d** This Licence constitutes the entire agreement between the parties with respect to the Work licensed here. There are no understandings, agreements or representations with respect to the Work not specified here. Licensor shall not be bound by any additional provisions that may appear in any communication from You. This Licence may not be modified without the mutual written agreement of DEMOS and You.